DREAMING IN TURTLE

Also by Peter Laufer

Organic: A Journalist's Quest to Discover the Truth behind Food Labeling

The Elusive State of Jefferson: A Journey Through the 51st State

No Animals Were Harmed:
The Controversial Line between Entertainment and Abuse

Forbidden Creatures: Inside the World of Animal Smuggling and Exotic Pets

The Dangerous World of Butterflies:
The Startling Subculture of Criminals, Collectors and Conservationists

Neon Nevada (with Sheila Swan Laufer)

¡Calexico! True Lives of the Borderlands

Hope Is a Tattered Flag (with Markos Kounalakis)

Mission Rejected: U.S. Soldiers Who Say No to Iraq

Wetback Nation: The Case for Opening the Mexican-American Border

Exodus to Berlin: The Return of the Jews to Germany

Highlights of a Lowlife: The Autobiography of Milan Melvin (editor)

Shock and Awe: Responses to War (editor)

¡See You Later, Amigo! An American Border Tale
(illustrated by Susan L. Roth)

Made in Mexico/Hecho en México (illustrated by Susan L. Roth)

Wireless Etiquette: A Guide to the Changing World of Instant Communication

Safety and Security for Women Who Travel (with Sheila Swan Laufer)

Inside Talk Radio: America's Voice or Just Hot Air

When Hollywood Was Fun: Snapshots of an Era (with Gene Lester)

A Question of Consent: Innocence and Complicity in the Glen Ridge Rape Case

Nightmare Abroad: Stories of Americans Imprisoned in Foreign Lands

Iron Curtain Rising:
A Personal Journey Through the Changing Landscape of Eastern Europe

DREAMING IN TURTLE

A Journey Through the Passion, Profit, and Peril
of Our Most Coveted Prehistoric Creatures

PETER LAUFER, Ph.D.

FOREWORD BY
RICHARD BRANSON

ST. MARTIN'S PRESS ❧ NEW YORK

www.stmartins.com

Library of Congress Cataloging-in-Publication Data

Names: Laufer, Peter, 1950– author.
Title: Dreaming in turtle : a journey through the passion, profit, and peril of our most coveted prehistoric creatures / Peter Laufer ; foreword by Richard Branson.
Description: First edition. | New York : St. Martin's Press, 2018 | Includes bibliographical references.
Identifiers: LCCN 2018026430 | ISBN 9781250128096 (hardcover) | ISBN 9781250128102 (ebook)
Subjects: LCSH: Turtles—Effect of human beings on.
Classification: LCC QL666.C5 L38 2018 | DDC 597.92—dc23
LC record available at https://lccn.loc.gov/2018026430

Our books may be purchased in bulk for promotional, educational, or business use. Please contact your local bookseller or the Macmillan Corporate and Premium Sales Department at 1-800-221-7945, extension 5442, or by email at MacmillanSpecialMarkets@macmillan.com.

First Edition: November 2018

10 9 8 7 6 5 4 3 2 1

As always
with love
for Sheila

CONTENTS

FOREWORD

Human beings should never be responsible for allowing a species to disappear from the Earth. We must all do whatever we can to keep that from happening.

On Necker, my island in the Caribbean, I suspect we have more animal species than on any other island on the planet. When species are in peril, a few should be kept in captivity to assure that the species continues to exist. Captivity for animals on Necker is not like captivity in a zoo. We make sure the animals enjoy plenty of room to roam. I feel like Dr. Doolittle sometimes, caretaker for so many happy critters, especially the day I watched one of our lemurs feed fruit to a couple of hungry giant tortoises.

The tortoises are among my favorites of the animals living with me on Necker. I was thrilled when we discovered our first Indian Ocean Aldabra giant tortoise egg on the island—another sign that our chelonian community is self-sustaining. In addition to the giant Aldabras, we provide a safe haven for critically endangered Burmese star tortoises and Burmese black mountain tortoises, along with such a comfortable home for red-footed tortoises that they are laying eggs and the eggs are hatching.

As I wrote in my book *The Virgin Way*, I favor seeking what

I call serious fun, and I enjoy doing good while doing well. That's why I appreciate the reporting that author Peter Laufer engaged in during the years he researched *Dreaming in Turtle*. There are plenty of good books about turtle biology and tortoise evolution. What makes *Dreaming in Turtle* a unique read is that it chronicles the serious fun experienced by the dedicated and disparate community of humans who have chosen to intertwine their lives with tortoises and turtles.

Of course there are scofflaws and other villains in the tortoise and turtle subculture. The poachers and smugglers and traders who prey on the animals are identified in these pages; it's important to know all the players—bad and good. But it's fascinating and reassuring to meet the motivated wildlife police struggling to save threatened and endangered species, and the hardworking fellow conservationists—many of whom I've crossed paths with as a result of our efforts on Necker to provide a safe haven. Their stories come alive as the heroes of *Dreaming in Turtle*, populating with unforgettable personalities this comprehensive and compelling survey of the status of these mesmerizing animals.

Listening intently to everyone with an opinion is another lesson I teach based on my business experiences. Latent herpetologist and swashbuckling journalist Laufer follows that rule in *Dreaming in Turtle* as he travels the world developing his own passion for tortoises and turtles while discovering their ubiquitous presence in all our lives. From Santeria priests in Cuba who sacrifice turtles for their clients to dedicated volunteers risking their lives to protect incubating eggs on Costa Rican beaches, from medicine men in Gabon who grind up tortoise shells to create hemorrhoid cure potions and Cajun cooks in Louisiana brewing turtle soup to undercover agents running risky sting operations in underground Southeast Asian markets teeming with endangered turtles and tortoises, the following chapters bring to life a fascinating cast of characters interacting with turtles and tortoises.

Dreaming in Turtle tells a cautionary tale of imminent

extinction—of what happens when timeless allure combines with illicit markets. But this is a story that can result in a happy ending. The book serves as a call to action because we all can, in our own fashion, help create for animals worldwide sanctuaries like those on Necker. And we must.

—RICHARD BRANSON, 2018

DREAMING IN TURTLE

PROLOGUE

OPERATING INSTRUCTIONS
FOR THE JOURNALIST

The Santeria priest and I are sitting on the floor in his Havana altar room. On a table across from where his last client left her cash (after the chicken sacrifice on her behalf) is a Buddha and a bust of Mary holding Jesus. Seven glasses of water are lined up on the white-sheet tablecloth along with an assortment of other icons: a portrait of a woman holding a sword and a jeweled goblet, a crinkled tract about Christ, a string of beads—one end dipped into one of the water glasses—and a lone cigarette. Changó—the lord of fire and lightning—is represented by a carved wooden object that looks like an urn, a bowl of squash and apples at its base, at its side a wooden statue of a monk holding a cross and wrapped in a string of beads.

"The *jicotea*," Babalawo Martinez instructs me about turtles, "thinks a lot. That's why he is very slow. The *jicotea* is always thinking." He tells me he feels connected to turtles. "The *jicotea* has lots of secrets, secrets given to him by the Supreme Being."

"For example?" I ask, and for the first time that evening Martinez balks at a question.

"The *babalawo* has a lot of secrets, too," he says about himself, the priest. "And the *babalawo* would be punished by the spirits of the dead and the orichas if he revealed them." But before we part he offers me the Santeria version of the prayer sung while Cuban priests cut the heads off sacrificed turtles; it translates to "Giving the blood to oricha who needs it."

This priest, Babalawo Martinez, is wildly popular with believers. As one of his followers told me, "He solves a lot of problems, and people see the results."

We stand and shake hands. "If you have any problem we can talk again," he says. At the time I assume he is referring to follow-up questions for my research inquiries. But as he says good-bye, he asks, "Do you have a turtle in your house?"

"Not yet," I answer.

"You must," he prescribes. "For the good energy and for your health."

I amble toward my waiting purple 1955 Oldsmobile Rocket 88 taxi, with its noisy Chinese bus diesel engine idling and fuzzy dice hanging from the rearview mirror, looking back at Sergio Martinez's crumbling apartment block. He stands in the doorway under the strand of dried grass that indicates to a needy public that a *babalawo* practices in the house, a lit cigarette in his hand. We exchange a waved adios.

Seems a bad idea to ignore an order from a Santeria priest. Soon after our Cuban encounter I identify the turtle to accompany me on my continuing chelonian journey. He? she? (I've yet to learn) is a rescue, picked up off the streets of Phoenix, Arizona, by a Good Samaritan who found a temporary home for it with Matt Frankel, the Turtle Conservancy board member who houses several dozen mostly endangered turtles and tortoises on his foundation's acreage outside Prescott. When I express concern about the responsibility of caring for my turtle, Frankel assuages my worry with this assessment: "They do wonderfully if they're neglected." My kind of housemate!

But I do expect, or at least hope, to develop a bond with my new companion. "As time passes, the two of you will get to know each other's personalities," I'm told repeatedly by turtle aficionados. Personality? A turtle? Yes, more than one source promises that turtles and tortoises show identifiable individuality, expressed, for example, by how one might eat out of a companion's hands, and whether it hides in its shell or reaches out for a caress. And apparently they do, eventually, recognize their human mates. Not that my Sonoran desert box turtle will necessarily be in a hurry to bond if I don't make the first moves. "Reptiles just slow down when there's nothing to do and they wait for the time to pass," reports its current caretaker, Frankel.

How will this new friend get to my home in Oregon from a thousand miles distant in Arizona? Ah, another example of the intriguing chelonian subculture. My turtle is scheduled to arrive via overnight express. Not just FedEx or UPS, but in a Styrofoam container filled with newspaper for it to cuddle against, delivered to my door by a company that contracts with those express services and that's found its own niche in the marketplace, a company named Ship Your Reptiles.

Come with me on my quest to document and understand obsessions surrounding turtles and tortoises. As I travel from the giant and infamous Jakarta animal market in Indonesia to a Santeria sacrifice ceremony in Cuba, from a sea turtle nesting beach in Gabon to the turtle-filled bayous of Louisiana, join me and meet a wild cast of characters, including a motley collection of outlaws and their customers—from poachers playing cat and mouse with game wardens, to those obsessive collectors seeking critically endangered species for their illicit collections, to patients desperate for cures and convinced successful treatment only comes from turtle parts, to gourmands ready to break laws for a unique meal.

Looking every bit as prehistoric as they are, turtles and tortoises have lumbered on our land since the age of the dinosaurs. Slow-moving and egg-laying, some live in water and some on land. Some eat meat, using their sharp beaks and strong jaws to slice through prey, while others subsist solely on plants.

These creatures all share a common feature: their bodies are encased in a protective shell that comes in a panoply of shapes and sizes and colors. Many of these species can pull their necks and heads into their shells, which are usually durable enough to protect them from the bite of predators. Their shell rings can hint to us of their age and their shapes telegraph whether they are swimmers or walkers—as do their limbs: some have feet, others flippers. They seem to enjoy good night vision, and tests suggest they're equipped with remarkably long memories. Some live well past a century. For as long as humans have been around, these animals and their eggs have been prized as medicine, religious and pagan talismans, food, decoration, pets, fodder for tales, and so they are of great value to collectors and traders.

Because of their value, the strange and marvelous turtles and tortoises are among the most trafficked and most endangered animals alive today—animals threatened by human greed, pragmatism, and rationalization. The story told here stars turtles and tortoises, along with both shady and heroic human characters, in settings ranging from luxury redoubts to degraded habitats, during a time when the confluence of easy global trade, limited supply, and inexhaustible demand has accelerated the stress on chelonian species. The growth of the middle class in crowded China and Vietnam, where the turtle is particularly valued, accelerates this life-or-death drama. But Asia is not the sole marketplace hoovering up the reptiles. This is a tale not just of endangered turtles and tortoises but also one of overall human failings, frailties, and vulnerabilities—all punctuated by optimistic hope for change fueled by dedicated champions of chelonians.

There is an existential threat to the world's endangered turtle and tortoise populations, populations that serve as indicator species for our own human survival. If Earth's environment cannot continue to support a stoic animal that's been thriving on the planet since the time of the dinosaurs, the rest of the animal kingdom must be considered in jeopardy and we humans—animals ourselves—should worry.

I research, report, and write about the relationships between humans and other animals. *The Dangerous World of Butterflies* is an investigation into the little-known enclaves of high-dollar international butterfly smuggling. *Forbidden Creatures* is a study of people who live with dangerous and endangered animals such as chimpanzees and tigers—documenting stories from seemingly happy interspecies families to ghastly tragedies. *No Animals Were Harmed* seeks to locate the frontier where animal use becomes animal abuse.

While researching the butterfly book I learned of the Philippine forest turtle, an animal thought to be extinct until a colony of them was located in an isolated island outback. The discovery resulted in frenzied collectors paying as much as $2,500 a specimen to poachers, who in turn paid as little as $50 to the locals who captured them—a precious $50 in a place where feeding a family is an economic challenge. The renewed trade again pushed the turtle toward extinction, and its start-stop survival story fueled my intrigue. The Philippine forest turtle led me to begin seeing the lives of turtles and tortoises as a metaphor for our own struggles to survive.

The turtle is not the typical charismatic poster-child animal—it's no cuddly koala, dog-faced harbor seal, or anthropomorphic polar bear—but it is rooted in many cultural identities. The turtle figures in Native American creation myths, and throughout history across cultures worldwide it also symbolizes wisdom, fertility, and long life.

Come along on my journey as I report little-known smuggling, document endangered species on the brink of extinction, and celebrate turtles and tortoises. *Dreaming in Turtle* is a love song to these magical, mystical, and mythological creatures. And it is a call to action.

In the waiting room at Trackside Auto Repair in Eugene I'm sitting in a worn and dusty chair while my old Volvo is repaired. Trackside is an informal shop with overtones of the 1960s on display: a framed illustration of Jerry Garcia, Rick—the grizzled mechanic with his ponytail sticking out the back of his baseball cap—and two junkyard dogs. I read. The little mutt (whose name I later learn is Axle), with no encouragement from me, jumps up and makes himself at home on my lap. What is his motivation? A pat or a scratch or a treat? Whatever. But there is no question in my mind that Axle made a decision. He chose to engage me. Will my turtle be so animated? I doubt it, but I am getting anxious to meet him (or her), anxious for the relationship to begin.

I've been thinking of turtle names. Fred comes to mind. Fred seems a solid identifier. And maybe unique for a turtle. Plus it works no matter the sex. Another candidate was Speedy, but that seems clichéd. When I ask my turtle benefactor, Matt Frankel, about its name, his response is dismissive. "It has no name," he reports. "These are not animals that react to names." Maybe, but I'm naming mine.

Our last cat was Schrödinger, a Himalayan flame-tip stray who never completely recovered from the psychological damage he had suffered. The poor cat arrived on our doorstep bruised with lacerations around his ribs and his whiskers snipped short. Schrödinger might fall asleep on my lap, enjoy pets and pats, and follow me up the street to the mailbox. But periodically and for no apparent reason he would become skittish and act as if we were strangers.

Will my turtle ever be anything but a stranger? Will Fred respond when I call out his (or her) name? Fred's scientific identification is *Terrapene ornata luteola*; he's called a box turtle because his hinged plastron allows him the capability to hide his extremities in his box.

Frankel is encouraging about the care and feeding of Fred. "You can keep it in an unheated garage," he says. "All you need is a four-foot by four-foot pen, a light, and a heating pad." Or, he says, I can keep Fred outside in the temperate Willamette Valley climate, "in a wire cage with a wire grid over the top to keep the turtle in and the predators out." Who are the likely predators? In my neighborhood raccoons and coyotes are prime suspects. "It can stay in the box most of its life," says Frankel, and only needs to be fed and watered every four days or so. But will Fred be happy confined to a box?

I'm both a tad nervous and somewhat excited about Ship Your Reptiles showing up at my front door with Fred inside his overnight Styrofoam home (I'm starting to think of Fred as a guy), cuddled up with shredded old newspapers. In fact, as the day of Fred's arrival comes closer, I realize I'm more than a little anxious. I'm concerned about the responsibility of keeping Fred alive and happy, the responsibility of keeping his pen clean and his water fresh and his food supply adequate. But I'm committed. Fred is en route.

Yet when I check back in with my turtle procurer to set up the shipment date, I seek reassurance. First I let Frankel know I've already named Fred, and he gives his enthusiastic approval. "That's perfect!" he tells me over the phone from Arizona. "He's Fredish. He's plain, not flashy. He's not a Francisco or a Federico. He's definitely a Fred." Funny how just giving him a name and then talking about him as Fred makes the upcoming relationship feel more intimate. We cover some of the basics again: he needs a room that stays at least 60 degrees so that he doesn't hibernate. (That would be no fun to watch for long and would make it hard for me to develop a relationship with him!) He needs a heat lamp to bask under that glows to 90 degrees and is at least two feet from him so Fred doesn't cook. He'll be happy with newspaper on his floor, newspaper that's changed when it gets soiled— about every two weeks, estimates Frankel. Feeding is two or three times a week: a half cup of fruit or vegetables, a spoonful of wet dog food, and a once-a-week live-earthworm treat.

"You can't screw up this thing," Frankel, an emergency room physician, reassures me. "It's not like a kid. Fred is idiot-proof. You can't break him." And he continues to make it clear that Fred is not a dog or a cat. "If he gets nothing to eat for two weeks, he'll be no worse for wear." Desert box turtles apparently can go months without eating and a couple of weeks without water. Frankel does instruct me to provide Fred with a water dish to soak in. He estimates Fred's age at between thirty and fifty years. His carapace is worn smooth in places from rubbing against rocks and brush he's crawled over and under, and there are some nicks in the shell, probably caused by tumbles from rocks. Frankel calls his weathered appearance interesting because it's evidence of his life experiences. "This is an old soul." We set the delivery date, and, finally, I find myself really looking forward to the arrival of this old soul, Fred.

ONE

THE MAJESTIC TURTLE

— Turtle Tales —

Everybody I encounter seems to have a turtle connection—from the stories of celebrity turtle fanciers (ex-49ers quarterback Colin Kaepernick, the musician Slash, CNN founder Ted Turner, comedian Sandra Bernhard, artist Julian Schnabel—the list is long and varied) to the experiences of Everyman. But the bond tying humans and turtles through history (and prehistory) is severely threatened because the turtle is in jeopardy.

"Everybody has a turtle story," I suggest over a casual drink with my friends Jim and Margaret. We are sitting on their deck overlooking Bodega Bay on the Sonoma County coast in California. It is an opening line I've been using since my turtle research started, and it never fails. At first Jim looked skeptical. "Everybody has a turtle story," I insist. "You have a turtle story."

He thinks for just a moment and then agrees. There was the little box turtle Jim kept in the backyard. They lost track of each other and Jim gave up looking. Five years later he was cleaning debris off the family swimming pool and, yes, there was the box turtle—and no longer so little. Jim fished the turtle out of the pool, and it promptly disappeared again and was never found.

Margaret's turtle story stars her grandmother at the age of

six or seven. She was down in the basement where the pet snapping turtle lived, and she must have been poking at it, because, according to family lore, the turtle grabbed her finger. Grannie screamed and screamed while trying to whip the turtle off her as it windmilled on her finger. Finally she succeeded at flinging it away. Back upstairs she asked why no one came to her aid. She was, after all, screaming.

"We thought you were singing," was the response.

Margaret's and Jim's turtle stories are kid stuff compared with the image my question conjured in Andrew DeVigal's memory.

"What's your turtle story?" I ask.

"I don't have a turtle story," he protests.

But I insist. "Everyone has a turtle story."

Now a colleague of mine at the University of Oregon, former *New York Times* journalist DeVigal thinks a minute more and then remembers that when he, his wife, and his children were on vacation in Southern California and the family was strolling along Venice Beach, "I saw a turtle walking alone with a hat on its back and it made me think about that powerful *Breaking Bad* scene."

Ah, yes, quite the nasty use of a tortoise in that show.

Everybody has a turtle story.

One of my students, Timothy Thompson, recounts a family tale starring his great-uncle Craig. Craig was five or six years old when he talked his parents into buying him a turtle for a pet. On the drive home, as Thompson recounts the family yarn, "Little Craig was holding his new pet when the turtle decided to withdraw into its shell. Craig thought this meant his turtle was dead. So he rolled down the window and chucked the poor turtle onto the freeway."

"I guess I'm just like a turtle that's hidin' underneath its horny shell," sang Janis Joplin in "Turtle Blues" on her 1968 album *Cheap Thrills.*

When I ask former U.S. poet laureate Kay Ryan for a turtle

story, she claims she has none. But she notes that she wrote a poem, "Turtle," and then recites a few lines.

"Why the turtle?" I ask.

"Because it's such a perfect emblem of having to go slowly and of clumsiness and of primitive sorts of movements," she says. "It's one of my favorite poems. It was written because I was so frustrated for so many years, which goes to show how valuable frustration is."

"Cat's Cradle!" says my son Michael when I bring turtles into our dinner conversation one night, and he quotes from a Newton Hoenikker letter at the start of the novel. The next day I find the passage in my collection of Vonnegut's works: "We all sat there in the car while Angela kept pushing the starter until the battery was dead. And then father spoke up. You know what he said? He said, 'I wonder about turtles.' 'What do you wonder about turtles?' Angela asked him. 'When they pull in their heads,' he said, 'do their spines buckle or contract?'"

Nope, neither. The neck vertebrae pull into a U shape when most turtles and tortoises yank their heads into their shells. Some species fold their neck sideways under the lip of their carapace.

Adventurer Yossi Ghinsberg chose not to eat a turtle when he was wandering alone, lost and starving in the Bolivian Amazon Basin. He lived to write a survival classic, *Lost in the Jungle*, and told his turtle story to the BBC for its *Survival Stories* radio series. "It was the first turtle I saw," he said about the potential meal that appeared in the jungle as he wandered, hoping for rescue. He imagined throwing a rock and it hitting the turtle. "And then it just looked at me. The moment our eyes met, something happened to me, and I actually asked his forgiveness and promised him I'm not going to hurt him. I let him go. I cannot explain it," he told the radio audience. "I had no problem to eat its flesh raw. But there was something that happened when the turtle looked at me. For a second I thought that we are the same thing in the same situation, and I just couldn't hurt it." This from a man who expressed no trouble killing, roasting, and eating monkeys during his ordeal.[1]

"Happy Together," a cheery ditty, was released in 1967. "I can't see me lovin' nobody but you for all my life," sang the Turtles. Not that the band or its members related to chelonians. Their name was just a gimmick, the band's cofounder, Mark Volman, tells me in 2016. "Our manager at the time recommended the name, thinking it would be misread as a British Invasion band," remembers Volman. "He thought that people would think that we were like the Beatles." The band was still touring when he and I talked; it was scheduled to play at the Kentucky State Fairgrounds in Louisville that evening. Volman divides his time: when he's not on the road with the Turtles, he teaches at Belmont University's Mike Curb College of Entertainment and Music Business in Nashville. "We couldn't come up with anything much better," he says about the band's name, "so we just took it." Despite the lack of connection to turtles, fans shower the band with turtle stuff. But as for Volman and his mates, "it was nothing to do with having any kind of relationship to the animal. We were looking for a name to get played on the radio and Turtles seemed like a good way to go."

The *flâneurs* of mid-1800s Paris strolled the streets for amusement and distraction as they studied the cityscape and its inhabitants. In his analysis of their languid pace, critic Walter Benjamin cites their use of turtles as a tool of their art. "Around 1840," Benjamin writes, "it was briefly fashionable to take turtles for a walk in the arcades. The *flâneurs* like to have the turtles set the pace for them. If they had their way," Benjamin concludes, "progress would have been obliged to accommodate itself to this pace."[2] I found the Benjamin essay in the University of Oregon stacks, a copy heavily annotated in pencil by a previous library client whose marginalia next to this passage is merely one word—an excited "Turtles!" I agree, the image of Parisian dandies out walking with their leashed turtles is intriguing, entertaining. Trouble is, aside from the undocumented

Benjamin reference, I find no record of the oddity. Nonetheless, ever since Walter Benjamin asserted it, the turtle-walking *flâneurs* show up as footnotes in all sorts of literature, from the academic to the popular.

Then there was Fluff, the Magic Turtle. Back when the Peter, Paul, and Mary dragon tune was popular, my friend Bob Simmons lived in Austin with a box turtle. "She kept going under the couch," Simmons remembers. "We'd lose her for days and we'd worry, 'What happened to Fluff?'" When Fluff would finally reappear, she'd be covered with dust bunnies. Simmons and his roommates came up with an easy fix. They epoxied a length of stiff wire to the top of Fluff's yellow and black carapace. The wire blocked Fluff from burrowing under furniture, and now she was easy to find as she roamed the apartment: flying from the wire was a hand-drawn American flag.

Stanley, the red-eared slider, was the adventurous turtle of another friend, George Papagiannis. He and I talked turtle over a bottle of wine in his Paris flat. His boyhood home was a Manhattan apartment on the fifteenth floor of a skyscraper with a strict no dogs or cats rule. But turtles were allowed, and a series of sliders lived in George's plastic oval turtle home with a green plastic palm tree and ramp from the island to the pool. George watched them and fed them by hand. The turtles died. His mother orchestrated a moment of mourning before they were, of course, flushed down the toilet. And they were replaced, one slider after another. But Stanley was special. He, claims George, had a distinct personality.

"Stanley was a standout because he stood up." Perched on his hind legs he routinely attempted to pull himself over the plastic walls of his enclosure. "He would stand against that clear plastic, trying." George still is convinced decades later that "Stanley wanted to figure out what was on the other side. For him, there was a bigger world out there."

The apartment featured a balcony overlooking the corner of Ninth Avenue and 23rd Street. When the weather was pleasant,

George would take Stanley outside to bask in the sunshine from the confines of a cardboard Bloomingdale's shirt box. Stanley would stand on his hind legs, scratching at the walls of the box, trying to see over its lip. One fine day, George went back out on the balcony to collect Stanley, but he was not in the box and he was nowhere to be found on the balcony. Curious Stanley had managed to get himself out of the box. "The place that he wanted to be was on the other side," said George, who was convinced his favorite turtle had taken a fifteen-floor flyer to the ground. "I was devastated. I couldn't believe it."

George raced down to the street and searched, but no Stanley. No remnants of Stanley. Stanley had simply disappeared. "The only thing I could think of was maybe a gust of wind had taken Stanley," George said. "My only comfort was that maybe Stanley landed on a taxi and now was living the dream, seeing all of New York from the hood of a taxicab."

Turtles fill our lore, from the classics to pop culture. Aesop's tortoise and hare teach us timeless lessons. Bert the Turtle fueled paranoia in the 1951 civil defense propaganda film *Duck and Cover.* The Teenage Mutant Ninja Turtles tried to entertain us while they peddled conspicuous consumption. Dr. Seuss created that megalomaniac Yertle the Turtle—the good doctor reminding us at the end of the tale, "And the turtles, of course . . . all the turtles are free, as turtles and, maybe, all creatures should be." The University of Maryland calls their sports teams the Terrapins (and nicknames them the Terps). The team's mascot is Testudo—Latin for tortoise. The Muppets Swedish chef attempts to make turtle soup in an early episode of *Sesame Street,* but the turtle fights back, pulls its head into its shell, then shoots at the confounded chef with a gun that replaces its head.

"It's turtles all the way down," explains what holds up our world. The origin of that theory is lost, but anthropologist Clifford Geertz tells it as an exchange between an Englishman and an Indian. "The Englishman," writes Geertz, "having been told that the world rested on a platform which rested on the back of an elephant which rested in turn on the back of a turtle, asked what did the turtle rest on? Another turtle. And that turtle? 'Ah, Sahib, after that it is turtles all the way down.'"[3]

These types of attributes, idiosyncrasies, and tales show the style and stuff that make turtles and tortoises unique characters of the animal kingdom and help create their special allure throughout human experience.

My first encounter with a turtle mirrors that of many grammar school boys. I can't remember where I got it. Nor do I know for sure what type of turtle it was. But it was likely a red-eared slider—the most popular pet turtle in America. It lived in a clear plastic tub punctuated, if memory serves, by a green plastic palm tree—the ubiquitous pet turtle habitat. It could swim around in water at the bottom of its quarters or it could crawl up its plastic ramp to bask on its plastic island under its plastic palm. We failed to become close before it succumbed to I do not know what. And, as I recall, it left the house in the manner of so many of its cousins: the poor little guy was flushed down the toilet—like Stanley, he was that small.

Turtles, capturing the human imagination, have represented longevity, fertility, strength, durability, and stamina to people worldwide. Our fascination with turtles too often has been at their peril. They are undemanding, if messy, pets. In religious ceremonies spanning the globe turtles are revered as deities or intermediaries to the gods. Before laws protected them, we routinely used their shells for guitar picks, combs, eyeglass frames, and jewelry. Turtle parts are still used—often illicitly—in traditional medicine, frequently as a treatment for sexual dysfunction. Turtle soup is common on menus; its meat is stewed and its eggs are mixed into elixirs.

As their populations continue to drop turtles have never been more valuable in the marketplace. Critically endangered species command high prices on the black market. Costa Rican businessmen travel from San José to the Pacific Coast to drink salsa infused with sea turtle eggs in hopes the potion revitalizes their sex drive. The upper middle classes in Asia are indulging a renewed (and frequently illegal) interest in turtles for pets, as ornaments, for food, and as ingredients in traditional Chinese medicine. Rare Yunnan box turtles list clandestinely for as much as $200,000. The same international trade routes that are used to run guns and drugs are also used to transport turtles illegally from their natural habitats to buyers in distant cities. Despite national laws and international treaties protecting the animals, the likelihood that turtle smugglers will suffer arrest is minimal and penalties are paltry compared with the profits.

To some skittish observers, turtles sure do not look as embraceable as koala bears. They may not be as easy to anthropomorphize as a chimpanzee. They do not tempt a bloody death like a pet tiger. But turtles tell great stories. Throwbacks to dinosaurs! Enviable longevity! As valuable to smugglers as drugs and guns! What's not captivating about turtles and tortoises?

The barrier that exists for many of us to identify with reptiles often influences their care and feeding. It never would have occurred to me as a boy that my "pet" turtle needed something more than a plastic container for its home, just as it never bothered me much when it was flushed down the toilet. Perhaps because reptiles don't look like us or come when they're called, they can seem disposable and replaceable—and without personality. It's easy to accept a double standard: relative luxury for cats and dogs, less than humane lodgings for snakes and lizards and turtles.

Lonesome George, the aged bachelor giant of the Galápagos, was much longer lived than my pet turtle. Old George, the last known Pinta Island tortoise (*Chelonoidis abingdonii*) on Earth, drew hordes of tourists, who came gawking at his hulking wrinkled self—a poster boy for endangered species protection. George stared back at the curious from his Darwin Station pen until he finally succumbed to old age at an estimated one hundred years. Fancy taxidermy performed by the American Museum of Natural History preserved this icon of extinction.[4] Now the curious can view George as an exhibit on the Galápagos, where he's polished and posed like a *Vogue* runway model.

Other Galápagos tortoises lumber around in their ample enclosure at the Turtle Conservancy's Behler Chelonian Center in Southern California, adjacent to the picnic table where I'm meeting with Peter Paul van Dijk. In addition to his role as director of Field Conservation Programs at the conservancy, van Dijk engages in important turtle research and conservation work with other organizations. He is the lead author of the annual *Checklist of Turtles of the World* and director of the tortoise and freshwater turtle biodiversity program at Conservation International. "I'm just one of many," he tells me with modesty when I call him one of the world's leading turtle experts, modesty accentuated by his measured responses to questions in a voice tinged with remnants of his Dutch accent. His introductory story is familiar. His parents acquiesced to his childhood request for red-eared sliders (*Trachemys scripta elegans*). It's been turtles for van Dijk ever since, from his PhD field research among turtles in their Thai habitat to his current assignments.

"They're living, walking dinosaurs," says van Dijk, explaining what he finds compelling about his life's work. "They're just so clumsy, so unrealistic, so improbable, and yet they are a success story." He marvels at how they survived when dinosaurs did not, how they thrived as climates and oceans and continents shifted. He calls them "ultimate survivors" if left alone. "Once they mature, like those"—he gestures toward the gawky, giant Galápagos tortoises grazing across the path from us—"they are pretty well indestructible and will go on for decades." Barring accidents, or a lucky and strong predator like a jaguar, they live long lives, "and they don't go reproductively senile." They can breed as long as they can breathe. Getting to adulthood is the trick: the eggs and the hatchlings and the young suffer massive attrition. Habitat loss and human lust for all things turtle combine to create existential crises for those adult turtles that do manage to beat the odds and mature.

Van Dijk celebrates the personalities he found in the turtles he used to keep. "I did connect with individual animals. They recognized me versus other people. I recognized them individually. I knew their preferences. I knew their temperament. There is very much a connection like with any living individual, whether it's a turtle, a dog, or a cat." He no longer keeps a turtle collection, and not only because his fieldwork often keeps him far from home. "I believe the best place for a wild animal is in the wild."

It makes sense that van Dijk as a turtle expert can differentiate between the animals in his care, but isn't it a stretch to determine that the reptiles recognized him? "There's more going on in those little brains than we give them credit for," he insists. "Their response to me walking into the greenhouse where they were living was different from my mother or a visitor walking in."

Maybe because he was recognized as a source of food and not as an individual? And what does he think is going in those little brains? The answer is quick and glib, accompanied by a sly smile.

"We don't know. We don't speak turtle yet."

Pressed to speculate, van Dijk lists basic instincts as turtle brain priorities: survival, food, and reproduction. But critical thinking, contemplation, musing, daydreaming—surely he doesn't attribute such behavior to our turtle friends.

"I wouldn't exclude that possibility," is the surprising answer from the scientist.

"Every so often they're deciding to go this way rather than to go that way. Why do they go this way? Is it purely random? I don't think so. There is some consideration going on in their heads that we're not even scratching the surface of." We've yet to figure out what that is, he theorizes, because we're not paying enough attention.

— *Dinosaurs Among Us* —

Contemporary turtles and tortoises belong to the order Chelonia and evolved toward their current status during the Jurassic period, coexisting with dinosaurs. The earliest turtle-esque reptile yet discovered—the ur-turtle that most matches its contemporary relatives—is *Proganochelys quenstedti,* a Triassic fellow who lived in Germany south of Stuttgart give or take 210 million years ago, as did kin on the other side of our world in what is now Thailand. There were some distant cousins evolving before *Proganochelys,* but they lacked the full package: a carapace on the back and a plastron protecting the belly—those shell halves that enclose and protect the turtle body. *Proganochelys* (the name means "before chelonians") carried some baggage missing from today's turtles, most noticeably, a club for a tail and armor on its legs. The club may have been a weapon or simply a defensive obstacle to predators.[5] The leg armor, along with the spikes on its neck and head, were for protection because *P. quenstedti* predated our contemporary turtles' unique ability to yank head and extremities into their shells for safekeeping.[6] Or so was the longtime prevailing belief.

It's Easter Monday in the Swabian Alps and almost everything is closed for the holiday—but not the museum in sleepy little Trossingen, a village just over an hour's drive south of Stuttgart. *Proganochelys* was unearthed in a gully here, and one of the fossilized ur-turtles is now housed here on its back in a glass case, among replica skeletons of the dinosaurs it roamed with.

The museum's volunteer curator, Volker Neipp, is opening the case for me. I want to commune hands-on with the oldest stem turtle yet discovered, the extinct immediate precursor of contemporary turtles.

"To touch her is special," he says about *Proganochelys.* "How long are you and I going to live? This girl lived before what we even think about. Imagine what she has seen."

Curator Neipp casually swings the glass door open and invites me to time-travel back 210 million years and touch the *Proganochelys* lying immobile on her back, two claws grasping at the air, dragonlike spikes on the edge of her carapace protecting her (and making her look so much the distant cousin of alligator snapping turtles). I reach into the case to grab a claw and hold it. "Careful!" calls out Neipp. "Don't break it!"

We hold hands a moment before we (or at least I) say good-bye; with that touch somehow I feel we are bridging her era and ours. I leave for another appointment while she spends her eternity there, both a link to Fred and another reminder that extinction is forever.

Up the road in Tübingen, Eberhard Karls University paleontologist Ingmar Werneburg, a *Proganochelys* specialist, is making news. He's confident fellow scientists erred in their analysis and that his models prove his hypothesis that the stem turtle could retract its head and extremities, making it an even closer relative to Fred. "The initial evolution of neck retraction," he and his colleagues assert, "occurred in a near synchrony with the origin of the turtle shell as a place to hide the unprotected neck."[7] But unless fossilized *Proganochelys* remains are found with legs and head pulled into the shell, the theory will remain just that.

Turtles, says Werneburg when we meet, help people understand evolution. "If they see a turtle, they see evolution. The world changed but the turtle did not." Ever the scientist, researcher Werneburg also displays a poet's approach to his field of study. "Sometimes turtles offer a little smile," he reports, "a smile that projects the wisdom of the earth."

Not that there aren't turtles today that can't do the head-in-shell maneuver. I met one—appropriately named the big-headed (*Platysternon megacephalum*)—when I ventured into that endangered turtle's Hong Kong range. There's simply no space in its shell for its big head; its bone-hard skull provides compensatory armor. Snapping turtles and alligator snapping turtles cannot retract their heads and legs into their shells. The aggressive snappers lure and stalk prey, counting on their nasty dispositions and vigorous sharp bite to combat enemies. With their armored legs and dragon-spike shells, snappers look closely related to *Proganochelys*. Today's sea turtles cannot retract their heads and legs either; they need their big flippers for efficient swimming and evolved with a streamlined body lacking adequate space to accommodate retracted extremities. The leatherback sea turtle and various softshell turtles—as their names suggest—do not sport hard *Proganochelys*-legacy carapaces.

Turtles live in the water, tortoises on the land, and terrapins commute between the two. They all bite with their beaks and chew without the help of teeth. The shell on the turtle and tortoise back—that trademark carapace—can be high-domed or all but flat and everything in between, depending on what a species needs in order for it to fit where it lives and protect against whatever predators may chase it. The carapace is the backbone and ribs spread out, fused together, and filled in solid by bony scutes that create the intriguing surface patterns prized by turtle collectors. Females stash sperm after they copulate, allowing them to lay fertile egg clutch after egg clutch without needing to engage another fellow. And the sex of their offspring is determined by the temperature where their eggs incubate: the hotties are female, the cool are male.[8] Hence climate change influences the male-female ratio.

Turtles and tortoises are as varied as they are extraordinary. The critically endangered Australian white-throated snapper sucks in oxygen via the same rear-end opening—its cloaca—that it uses for excretion, mating, and laying its eggs. The Chinese softshell turtle voids urine through its mouth. The North American alligator snapping turtle sticks out a wormlike tongue appendage, tricking fish to follow it into its mouth. The South American red-footed tortoise clucks like a chicken. The Eastern musk turtle scares off attackers by releasing the odor that earned it the nickname "Stinkpot." Turtles can be tiny three-inch, five-ounce Cape turtles or huge two-ton mammoth sea turtles. There are carnivore turtles that steal bait off hooks of fishermen and alligator snapping turtles that eat common snappers. There are strict vegetarian turtles. Say hello to the critically endangered angonokas, more commonly known as ploughshare turtles because of the prehistoric-looking ploughs jutting from under their little heads like the cowcatcher on a steam engine. Their home range, as lawless as any in the Wild West, is in Madagascar, and they are being decimated by illegal poaching: each ploughshare is worth as much as $50,000 on the black market—ten times what a Philippine forest turtle brings to a smuggler. Captive breeding, designed to save ploughshares, fuels the illegal trade as new generations, vulnerable to poachers, are returned to the wild. Soviet space scientists sent a pair of turtles around the moon, the first Earth animals up there, predating by some six months Neil Armstrong's famous step on the lunar surface. And then there's Jonathan, a turtle still living as of this writing at the British governor's quarters on the island of St. Helena, and who, at 183 years, may be the oldest living creature on Earth.

• Fred Arrives! •

I'm rushing out for an appointment, grab my lunch and pull open the front door. There, sitting on the doormat, is a FedEx package with LIVE ANIMALS scrawled on it. It's a cold, rainy day. "Fred's here!" I call out to Sheila, and since I'm running late for work I leave the box for her to deal with. By nightfall I'm not yet home and a worried Sheila decides to open the carton. Inside is Styrofoam popcorn protecting a plastic storage container, and inside that is a cloth bag. "I was relieved to see a little movement in the cloth sack he was in," she wrote in an email message to me. "He also moved again when I talked to him." She cut the rubber band that secured the bag and placed it on the living room floor. "His head came right out and he looked at me. 'Hi, Fred,' I greeted him. He began to walk fast, went into a corner and stayed there. He is not a slow turtle." No question that by the time I get home Sheila's bonded with Fred. She talks to him, talks about him, worries about his adjustment to a new home, gives him water, lettuce shreds, and diced apple. And who knows, Fred may feel the same about her.

He looks the size of half a small cantaloupe. I pick Fred up and place him—gently, of course—in his house. After dinner Sheila checks on him again and I overhear her lullabying him with the same sweet, soft contralto she used with her human babies.

TWO

THE TIMELESS ALLURE

— Ancient Oracles —

Tucked away on display in an alcove on the third floor in San Francisco's Asian Art Museum is a turtle bone fashioned into an oracle. It's just a fragment of turtle an inch or so on each side, displayed with a few larger oracle bones listed as "probably cattle bone." Two characters are scratched or carved into the turtle bone. One looks like half an arrow bisecting an upright line with a flag above the intersection, and the other looks like Neptune's trident. The relic dates from sometime in the Shang dynasty (circa 1600–1050 BCE).

The turtle bone was collected near Anyang, in Henan Province, and it is one of more than forty-five thousand found and cataloged by archaeologists. "Most foretell births, deaths, rainfall, good harvests, the outcome of hunts and battles, and the meanings of dreams," says the museum label.

It seems a propitious stopover for my turtle journey, that message: looking for the meanings of dreams, along with the longevity of the marked remnant and its previous incarnation as a living turtle's bone from the dawn of recorded history. Over three thousand years later across the globe in the dark corner of a museum vitrine, its current home is thousands of miles from where it was used as a type of journalism. I take a last look at the oracle and say good-bye. I do not know the meaning of the engraved semaphore and pitchfork, but that ages-old elusiveness reminds me not to take too seriously my own daily passing irritations, like when the parking meter in front of the museum clicks to EXPIRE and a meter maid slaps a $76 San Francisco ticket on my old Volvo's windshield.

In ancient Greece lyres were made using tortoise shells as resonators.[1] The Mayan god Pawahtún, who holds up the sky, takes the form of a human in a turtle shell (and often appears intoxicated, seeking the company of attractive women). In the *Egyptian Book of the Dead*, Thoth says to the sun god, Ra, about his enemy the turtle, "Behold, with strength I force my access to the sun disk. Truly, Ra lives! The turtle is dead."

Turtle allure connects our contemporary world with that of our ancient ancestors, just as turtle physiology connects Fred with the dinosaur era. In Vietnam, for example, an iconic turtle embodied nationhood. Called Cụ Rùa, meaning Great-Grandfather Turtle (despite being female), it was revered in Vietnam as a symbol of Vietnamese independence, the guardian of the sword that has protected Vietnam for centuries, a sword Cụ Rùa lent to a Vietnamese military hero who fought off invading armies from China. When the Yangtze giant softshell turtle died early in 2016, the nation grieved: the death of this 360-pound creature, found floating in Hanoi's Hoàn Kiếm Lake, portended bad luck.[2] Turtle biologists were just as upset as the Vietnamese people. Cụ Rùa was one of the last known specimens of the species *Rafetus swinhoei*. Turtle researchers knew of only three others when Cụ Rùa succumbed: one wild in a Vietnamese lake (perhaps a male), the others a male and female couple in a zoo in China.[3] No babies for them—yet. The male's penis was damaged in combat with another fellow (who died in the battle). Nonetheless, there still existed a sparkle of hope that sperm could be collected from the damaged guy and his partner's eggs artificially fertilized, or that the wild one is a male and he (or his sperm) could be introduced to the gal in the zoo—and that one or the other strategy might keep *Rafetus swinhoei* from going extinct.

"All thoughts of a turtle are turtle," wrote Ralph Waldo Emerson enigmatically in an 1854 journal entry. Creation myths of the Americas abound with turtles, varying according to region and tribe. The Great Spirit deposited the earth on the back of a giant turtle. Turtles symbolize the Earth. Turtle Island is embraced by some Native Americans as the name of their home continent. It's also the original name of the company that makes my Tofurkey breakfast sausage and the name of one of my favorite string quartets.

— *Turtle to Cure What Ails You?* —

"The Chinese definitely have a fascination with turtles," Heiner Fruehauf says. He knows. Fruehauf has spent much of his professional life studying China and traditional Chinese medicine both at American academic institutions and in China. Since his first fieldwork in China in 1982 he's returned every year to conduct further research. With a PhD in East Asian languages and civilizations from the University of Chicago, he teaches at the National University of Natural Medicine in Oregon's hipster capital, Portland, where he practices what he calls "classical Chinese medicine," dispensing compounds for treatment that are not available at your neighborhood Walgreens and Rite Aid—including treatments made with turtle parts.

Although the consensus of Western modern science refutes claims that turtle parts are curative, Fruehauf is one of the outliers. We meet in my hotel lobby. Fruehauf, his thinning white hair adrift, arrives with a satchel stuffed with a pile of extraordinarily fat books—all written in Chinese. "The Chinese are incredibly prolific catalogers of the material world," he says as he hefts one of the thick volumes out of his bag. All things that were known early in China's history, he says, are listed in the two-volume book he's showing me, along with an explanation of what purpose they serve. "This is called," and as he looks at the characters he translates the title into English, "'Ancient Thing Encyclopedia.'" This cataloging, he tells me, manifests itself in the herb world (he uses the word "herb" for anything—plant, animal, or mineral—that is used for its medicinal value). Next out of the satchel is a book that he says lists all the curative herbs in use a thousand years ago.

"Many of my Chinese teachers have turtles for pets," he notes, as he proceeds to explain why the turtle is a symbol in China for longevity. "There is this idea in Chinese medicine that you only have a given number of life force, and the more you run and work and get agitated, and also breathe and move around, the more you squander that life force and the shorter your life will be." The Chinese term for that life force is pronounced "chee," generally transliterated as "chi" or "qi" in English. "If you preserve your chi, and you use it sparingly and you breathe less and you have fewer ejaculations and you have less excitement and you work less, then you store your energy." Your chi. Cue the turtle. They live longer than other animals. "From an observational perspective the Chinese then say it's because the turtle never moves. So," says Fruehauf as he interprets the lore of traditional Chinese medicine, "that means it's a master of storing chi." Turtles are appreciated not just for their longevity but also for their wisdom. It's a wise critter that manages to last so long.

That mastery of chi storing—resulting in turtle longevity—sets the scene for the use of turtle parts as tonics. The concept is called "signature": if a plant or animal behaves or looks in a specific manner, making use of it as a remedy supposedly stimulates similar results in the patient. In the case of turtle as longevity tonic, Fruehauf explains, turtles live a long time because they contain a certain type of energy that—if consumed—will stretch the human lifespan. "The idea is," he says, "if I eat a tiger penis, I'm going to become all ferocious and powerful and potent. And if I eat a turtle, then I live particularly long and become wise." But beyond the turtle metaphors, Fruehauf claims that he finds actual results. "Because the turtle does live longer than other animals, when you use the actual substance from its body, it can give you some of the chemicals of longevity. That is the science of signature." Belief in such unproved "science" fuels the assault of desperate humans on other animals, many of them highly endangered, from tigers and their penises to rhinoceroses and their horns to turtle meat, shells, and eggs.

Superstitions are based on such signatures, agrees Fruehauf when I question the efficacy of such treatment, but, he insists, not just superstitions. "From a clinical perspective I'm a firm believer that the science of signatures works. Where it becomes superstitious is to believe you take a turtle so you live long." Traditional Chinese medicine classifies substances as yin and yang, explains Fruehauf. Yang tonics, such as those made of ginger, cinnamon, and horseradish, "make you feel warm and energetic." Yin tonics "lubricate the system and calm it down"; they're used for anxiety, insomnia, and the like. "The turtle is an animal that knows how to preserve its energy," Fruehauf says. They are yin, the opposite of the fast-running hare. "Imitating that spirit and behavior probably is doing much more for your body than eating actual turtle flesh."

We talk for a long time. Fruehauf's soft, gentle voice makes for easy listening as he goes into great detail about "life forces" and connections between European and Chinese folk beliefs. I appreciate the background but try to keep the focus of our chat on how he claims turtle works as medicine. "There are some people who don't live long because they have a yang deficiency. Giving them turtle would be contraindicated," he says, as he hauls another of the heavy tomes out of his bag and flips through it, showing me pictures of—well—apparently everything animal and mineral known to the Chinese centuries ago. "Not just turtle, but everything in nature is a medicine," he says is the message of the book. "The liver of the mouse, the tail of a mouse, the hair of the mouse, the tooth of the mouse. Everything is a medicine in nature."

Again I bring us back from mice to turtles, asking if Fruehauf considers turtles medicine. He laughs when he says yes. "The turtle is medicine, but so is everything else." It's the plastron, "the belly plate of the turtle," that's prized. It's boiled—he equates the process to making bone broth for the trendy paleo diet. One type of turtle, he says, is a treatment for osteoporosis because it can solidify bones. Another, a softshell variety, is used to treat cancer. "Modern researchers," he maintains, "say that there is something in the *bie jia*," the softshell turtle shell used in traditional Chinese medicine, "that prevents tumors, particularly malignant tumors, from absorbing iron, and therefore they starve to death. This turtle has a mass-softening effect."

"And you would treat a cancer patient with this turtle-based concoction?" I ask.

He answers, "Yes," in his soft, almost hypnotic monotone, but I'm not prepared for the addendum to his response. "Because I myself had cancer. I got cancer myself just when I got my Ph.D., and I changed careers to traditional Chinese medicine because I wanted to find a way to heal myself other than just going through the Western medical route." He says he is cancer-free these many years later after a combination of Western and Chinese treatments for his testicular cancer. "I never looked back and became a professor of Chinese medicine." He doesn't recall all the compounds the practitioners who treated him prescribed, but he figures the *bie jia* is a likely candidate, along with "lots of bizarre animal materials like a type of dung beetle I can remember having taken and some kind of toad that had bumps on its back that a particular doctor said looked like testicular swellings." But he quickly puts toads and beetles and turtles into perspective regarding his therapy. "If I say I owe my life to Chinese medicine, it's certainly not to any one ingredient. It's more the genius of thousands of years of being able to assess the medicinal quality of different parts of the natural world."

Or, thinks the skeptical Western journalist interviewing him, it was the effect of Western medicine, and the Chinese stuff was at best a placebo. Out for another show-and-tell comes another two-volume reference with about six thousand different herbs listed that are available to modern traditional Chinese medicine practitioners.

Heiner Fruehauf uses farmed turtles when he prescribes turtle treatment to patients at his Chinese medicine private practice in Oregon. "They are adequate for the purpose." He decries the taking of endangered species. "Some Chinese people get the idea that the wilder and the more rare and the bigger the turtle, the more tonic it must be." In addition to his medical practice, Fruehauf operates his own herb company. One compound he sells that's used for ovarian and prostate cancer (he's quick to add there is nothing written on the bottle advocating its use for cancer because it is an unregulated "supplement," and claiming specific health benefits would violate FDA rules and regulations) contains softshell turtle parts. He calls the compound effective, but he worries about sourcing the turtle from Chinese turtle farms, fearing toxicity from their excessive use of antibiotics. He would prefer using wild animals. "There is definitely no question from a clinical perspective that anything in the wild is vastly superior as a medicine to something that is factory-grown." And he hedges regarding the turtle ingredient being crucial for a cure, "because there are other materials" in the mix he sells. Fruehauf would prefer to forgo putting turtles in his supplement soup and he seeks alternatives. "There is no need to mess around with animal materials if plants work just as well."

But if a European- and American-trained academic embraces turtles for therapy, it's no wonder that so many Asians, especially those desperate for a cure, look back at their cultures' histories and anxiously seek turtles—common or endangered. The belief that wild is better depletes populations, as does urbanization. "It's just terrible to see the decimation of the wild places in China," says Fruehauf. "I'm just heartbroken seeing the destruction," and he ticks off a list of locales in China where turtles play a secondary role to so-called progress in the form of massive new cities. "They're building skyscrapers in the middle of rice fields and it's just terrible," says Fruehauf. And China's wild nature reserves are no safe house for turtles: hunting is rampant in the refuges, and China's strict laws forbidding it are all but unenforced.[4]

"Government officials who are supposed to penalize people for poaching themselves go out and poach," Fruehauf says. He uses a word in Chinese to describe the corruption he sees in contemporary China: *hòumén*, which is the Chinese term for "go in the back door." "China right now has an incredible ethical crisis," explains Fruehauf. "There's all this money around but no ethical values and no control. Now it's just money and what you can get away with. The disrespect for nature and animals is huge. When you go to a Chinese zoo, it's heartbreaking. They throw stones at the lion just to see what he'll do."

When it comes to animal abuse, humans know no borders.

• **Is Fred Okay?** •

The day after Fred arrives I fix him up with his heating pad and basking light, an overhead 50-watt Repti Tuff unit from the Zoo Med company that's designed to generate heat to make Fred happy in the Pacific Northwest winter, a halogen lamp that Zoo Med promises "makes animals' colors appear richer." His house, void of accouterments, worries Sheila. What is he going to eat, to drink, to do? And I'm late for work. But I dash off a quick email to Matt Frankel. "Fred survived the night," I tell him.

Frankel's response both makes fun of us and offers some solace for our concerns. "Survived the NIGHT?" he writes. "He's a turtle. He will survive the decade. Even the election." (We're less than a month from the 2016 presidential vote.) "That is part of their magic," he raves about turtles, "300 million years!!" I write back that Sheila loves Fred. "Glad he is in a good home," Frankel responds. But when I relate Sheila's concern that Fred's lonely in his house and tell Frankel of her scheme to cook some organic chicken for Fred instead of feeding him the dog food he recommended, Frankel is undone. "Turtles live alone. He would probably prefer the bugs off of four-day-old road kill to boiled chicken. Dog food is fine. Veggies over high protein. You can explain anthropomorphizing to Sheila. I am sure she knows. Turtles live long because they don't care. Note the period! A good lesson for humans." It's a lesson I look forward to learning when my schedule allows me some quality time with Fred. Meanwhile, Sheila offers Fred organic raspberries from Whole Foods and organic lettuce from the local Excelsior Farm.

He expresses zero interest in both.

— *Ancient Turtle Meals* —

Soon after I first moved to the high Nevada desert years ago, I was careening from Silver City to Dayton in an old truck with an old friend. Jim took a last swig of his beer and hurled the can—as was his custom—out the window. "Ah," he told me, satisfied with both the cold beer and the easy trash disposal, "the desert! God's garbage dump." Jim wasn't alone. Humans have been trashing our deserts since we arrived.

A couple of blocks from my office sits a white frame bungalow, headquarters for University of Oregon archaeologist Dennis Jenkins. He's famous for prowling the Wild West and finding evidence out in the desert that he says dates human habitation in those parts back thirteen thousand years ago—to the oldest human remains in the Americas. I wander over to his laboratory one day because I heard around campus that Jenkins likes to tell tales about his encounters with the desert tortoise, and the lessons they can teach us. "Tortoises are survivors," he says, impressed. "They have a hard life and they have adapted to it. I think that just speaks to us."

Back in the early 1980s, Dennis Jenkins practiced his science at Fort Irwin, the U.S. Army's desert training site in the Mojave Desert. "Because we're going to be fighting over oil for as far as we can see into the future," Jenkins recounts to me, "we prepare for conflicts in deserts." That training means Fort Irwin is home to a constant parade of soldiers and their equipment and, as Jenkins explains it with what I quickly learn is a trademark no-nonsense analysis of what he observes, "The army is destroying the environment. There's no getting around that."

The archaeologist leans back in his chair. A big silver Western-style belt buckle holds up his blue jeans. A fleece vest completes the cowboy look. One of the army's victims, he says with a refined drawl, is the desert tortoise. "What I know about tortoises is that they lose when they encounter a sixty-ton tank or even a thirteen-ton armored personnel carrier moving at forty miles an hour. They just get squished." But Jenkins looks at the tortoise's predators—intentional or accidental—with an archaeologist's timeline. "This has been going on for thousands of years."

Long before the U.S. Army existed, Native Americans were enjoying the desert tortoise. "We found tortoises that had been cooked when we were digging at Fort Irwin. In one of the rock shelters we were excavating we found probably eight tortoise shells, stacked inside of each other, that had been cooked. Somebody was heavily into eating tortoises." Jenkins dated the find at about a thousand years old. "They had broken open the bottom of the shell and eaten the tortoise right out of the shell—cooked it in the shell." The researchers ascertained that their find was leftovers from a dinner party because of the burned tortoise leg bones they found at the site, bones scarred with butcher marks. Tortoise was a good deal for the desert Indians. "It was free basic calories that you could catch. Once you caught them you could pack them around for as long as you wanted—a live source of protein waiting to be consumed at your leisure."

The archaeologists were surveying sites in jeopardy at Fort Irwin and found that tanks were just one contemporary tortoise enemy. "I remember distinctly—because I was shocked by it—finding a pile of baby tortoise shells. Each one had holes broken in the bottom and the baby had been consumed." A quick look around revealed a raven's nest above the pile. "This raven was feeding its babies baby tortoises as they emerged from the nest, as they were being hatched out." Under natural conditions, ravens eating baby tortoises should be no big deal for the tortoise population. "But when you add that to tortoises getting squished, pollution, people picking them—" Jenkins interrupts himself and takes me back to his Las Vegas boyhood.

"I had a pet tortoise for years. We kept him in the backyard. Or her—I don't know what it was." He laughs at not caring about such things when he was ten years old. It was about a foot long when he grabbed it in the Red Rock Canyon. His family was driving out into the desert for a picnic. "My dad said, 'There's a tortoise on the road!' We were concerned about it getting run over and we just took it home." The Jenkins family was just one of the thousands of other tourists who were lured by the slow-moving beasts. "We never thought about it. It was the 1960s. We fed it lettuce and things that must have been horrible for it." He laughs at the memory but with a glimmer in his eyes, a glimmer that suggests to me he really liked hanging around that tortoise. "I'm sure that it didn't get the proper nourishment that it needed. But it lasted for years." The family dug a hole in the backyard for its den. "Eventually it escaped from our yard. I don't know how. Somebody left the gate open or something."

"And that was that?" I ask.

"And that was that. It disappeared. I think somebody else took it home."

It's a Southern California tale as common as running off to Hollywood seeking stardom. Everybody has a turtle story.

"I love the desert," Jenkins tells me, dragging out the word "love" in a hoarse stage whisper. "There is something that calls to me from the desert. The desert is honest. It's exciting because it is somewhat unforgiving. You have to know what you're doing."

"And the tortoise is an example of that desert magic," I suggest.

"It is exactly an example of the spirit that calls to certain people who are desert rats, and I consider myself a desert rat."

"You kind of look like a desert rat," I say, quickly adding that I consider it a compliment. "You've got the twinkle in your eyes, the gray beard and the gray hair, and the leathery skin."

"Absolutely, the road map face. It comes from living in the desert."

Desert tortoises are still facing the same threats that Dennis Jenkins witnessed at Fort Irwin thirty years ago. Army training for combat in the Middle East continues to encroach on tortoise habitat, as does urbanization. And lately, a new competitor has arrived on the scene: huge Mojave solar farms. The tractors and trucks carving and scraping the landscape squish tortoises with the same efficiency as an APC or a sixty-ton tank.

— *Desert Tortoise Versus Solar Farms* —

One of the most massive of solar farms is the Ivanpah Solar Electric Generating System, a power plant built by the consortium called BrightSource with a price tag of well over $2 billion and hundreds of sprawling mirrors over a few thousand acres of public land on the California-Nevada border. Its enormity dominates the vista alongside Interstate 15 at the state line.

The desert tortoise (*Gopherus agassizii*) is listed as threatened on the Endangered Species List. The population is 10 percent of what is was in the 1950s. The hearty chelonian has been a resident of the Mojave and its triple-digit temperatures for millions of years—surviving the heat by digging a shelter deep underground. As part of the solar project deal, BrightSource promised to mitigate damage to the protected animal by relocating individuals found at the site. As an example of the stakes, consider the budget for the tortoise work: over $50 million. The

money is spent hiring biologists to move about seventy-five tortoises from the work site, care for them while they're in transit,
and relocate them. But indications are that it's $50 million down
a tortoise burrow, with the animals the losers and scores of private biologists reaping riches as if they were winning big at one
of the casinos in Primm, Nevada, two miles up the interstate.

I caught up with a former BrightSource contract biologist,
Mercy Vaughn, in Arizona. Experienced with threats to the
desert tortoise, she spent a year and a half working on the
Ivanpah project before she and the company divorced. Despite
her continuing worries about the tortoises that were in her care,
she remains convinced that such a solar power plant and the
threatened animal can coexist—just not necessarily under the
terms BrightSource has dictated. Rather than build solar farms
on pristine landscape like the Ivanpah Valley, land adminis-
tered by the Bureau of Land Management, she favors locating
them on terrain already compromised by previous exploitation: old agricultural fields and failed housing developments,
for example.

In its literature, the Oakland-based solar company calls its
mitigation efforts a success. "As a result of our tortoise care and
protection efforts," BrightSource claims, "many more healthy
adult tortoises will be returned to the Ivanpah Valley than would
have survived had the project not been built."

"Is that correct?" I ask Vaughn.

"You know," she begins, and then hesitates, explaining, "I'm
not at liberty to discuss a lot of details because I am under a legal
binding agreement with them that I can't talk about details of
the project."

"Sure," I coax her, "but we can talk conceptually. Are the turtles in the soup?"

"Conceptually," she embraced the word, "Desert tortoises do
not translocate well. If you are going to translocate a desert tortoise, translocating them within their home range is better
than taking them out of their home range." At Ivanpah, she told

me, some of the tortoises were taken out of their home range. Others stayed in their native neighborhoods.

But aside from whatever concern she may have about individual animals, biologist Vaughn worries about the survivability of the threatened species. "The fragmentation of the habitat is immense," she says. "Its impacts are profound," she says about the solar farm. "You're restructuring its entire system," she says about the tortoise's habitat, "by placing this obstacle that they can't penetrate in the middle of it." Movement patterns are altered, affecting not only the animals translocated but also those animals around the monster project. That means shelter, food, and water used by the tortoises since their time began are off limits to them. "You're taking away a whole set of resources for every individual animal." The effect, she is convinced, will radiate outward far from the project, "because you're pushing these animals out farther and farther so the ones that were living around the facility now are pushed into the home range of those that were living even farther away. How far out does that ring of effect go?" she wonders, and then answers her own query. "Who knows?"

The facility is fenced to keep adult tortoises out. But hatchlings can clamber through the mesh, back into the solar farm.

"This is all a giant biological experiment," she says about the desert solar industry.

"Does that scare you?" I ask.

"I've gotten to the point where I'm pretty fearless and cynical about it all, to be perfectly honest with you," she says, laughing as she adds, "I think we're screwed. I think we're masters of our own demise. Am I optimistic that we have the ability to change our course?" Again she answers herself, "No, not really. There's so much greed and desire for power, it would take altering our lifestyle and making people responsible on an individual basis for their own production of energy and the accounting of what they consume."

If she fears the world is coming to an end, why worry about

the wild desert tortoise, especially when so many of them are thriving as pets in private hands all over California?

"That's a good question," she says, telling me that out of respect for a fellow living creature she no longer believes trying to save the tortoise in the wild is a responsible act since so few survive when their home range is ravaged.

"That's a sobering conclusion," I say.

"It's very sobering for me," she agrees, "because I'm a bleeding heart. I'm somebody who has spent twenty-four years of my life as an advocate for this species. Yeah, this is incredibly sobering. I've shed a lot of tears about it over the years."

A couple of weeks later I was out on the edge of the Mojave talking with another one of the two hundred or so biologists who have been ordered silent unless their words are vetted by their employer, BrightSource. This one—thoroughly dissatisfied with the project—chose to talk freely, once I guaranteed anonymity.

The biologist doesn't fault BrightSource for the disaster the tortoises and the desert are experiencing, but rather the frailties of the Endangered Species Act and the clumsy bureaucracies at government agencies like the Bureau of Land Management (BLM) and the U.S. Fish and Wildlife Service (FWS). "They're the ones that should be held accountable here. BrightSource has done exactly what they've been told to do and have actually given us more money and more resources than required," the biologist says, pausing before adding cynically, "for PR probably." But the biologist does not see BrightSource as a villain. "They're just a company trying to make a living." The plant is not in a designated critical place for the tortoise, so a case cannot be made that the survival of the species is threatened by the project. Current regulations cannot stop construction. "The species is threatened and numbers have decreased dramatically," which the biologist attributes to habitat loss and the upper respiratory disease that the desert tortoise is prone to contract. Nonetheless, routine mitigation work on the project site carries emotional costs. "When you're walking in front of a bulldozer, crying and moving animals and cacti out of the way, it's hard to think that the project is a good thing."

Despite the biologist's attempts at appreciating all sides of the controversy, a conclusion is inescapable. "Translocation is a terrible idea. Everybody knows that. Everyone knows it doesn't work." The biologist cites the Fort Irwin expansion as an example of such failure: in 2008, over seven hundred tortoises were moved by the army, and scores died wandering around unknown territory—killed by hungry coyotes, disease, and traffic. And the biologist is not naïve about the power company's motivation. "They don't go about it saying, 'What's best for the environment?' They go about it saying, 'How can we make the most money?'" Nonetheless, "they really believe that what they are doing is good. They want to do the right thing." However, "I don't think the amount of energy we're getting from that site is worth what's happening to the desert. It's not that much."

Yet rather than spending millions and millions of dollars trying to save the few desert tortoises living in the Ivanpah Valley, the biologist figures, better the animals be ignored. "We should just leave all the tortoises. A bunch of them would get killed. Take that money and go spend it on preserving habitat elsewhere," because if all the solar farms proposed for the Mojave are built, this expert forecasts that the wild desert tortoise will cease to exist.

— *A Pet Desert Tortoise* —

I grew up in northern California and, until I went to the Mojave to check out their plight, I'd never seen a desert tortoise. But I heard the stories: the threatened species is a common sight in backyards from the Mexican border to Silicon Valley.

My first meeting was with a creature aptly named Torty. Torty was spotted looking lost and vulnerable on a Southern California road. Lily Hymen and her husband saw him while they were taking a walk, and they decided to rescue him. They brought him to their bungalow in an isolated valley not far from the coast and checked with neighbors "to see if anybody was missing a turtle." Nobody was, she told me, "so we kept him."

"What's the attraction?" I asked Lily (we were on a first-name basis as soon as she welcomed me into her home). "Why would you want to pick him up and bring him home?"

"He seemed to be lost," she smiled at the memory. "He could get run over. And we liked turtles."

That was over thirty-five years ago. For most of that time Torty lived in a cardboard box upstairs in her house except when he was hibernating. For that he favored a burrowlike brown paper bag on the porch. These days his digs are more luxurious; he's been upgraded to an enclosure in the side yard and a stone hut. Torty was hibernating when I met him, a brown lug of a guy with a carapace about a foot long, maybe longer. His head, hands, and feet were tucked in for the winter. To me he looked more like a satchel than a pet.

"He's a great pet," Lily insisted. "Six months of the year he's asleep and six months of the year he's eating." Low maintenance. "He recognizes my voice. When he's over on the other side of the pen I say, 'Hi, Torty,' and he comes over to me."

Lily calls Torty a charming animal. "I find him fascinating. I can't explain it. I don't know why." She speculates that perhaps it is his long life expectancy. Torty could live another thirty-five years, no problem. "They're very slow and quiet, and that's so wonderful," she told me as I left the two of them, "because everything these days is fast, fast, fast—you have Kindles and iPads and iPhones. I think it's affecting everybody's brains. And he just plods along."

— *Official Desert Tortoises* —

In an effort to keep track of the Tortys of the world, the state Department of Fish and Wildlife works with the California Turtle and Tortoise Club to offer abandoned desert tortoises to rescuers who wish to adopt them. Since desert tortoises are listed as threatened, it is technically illegal to own one. But thousands now live—illegally—in Southern California suburbia. Dave Friend is the club's executive chair. We meet at his Ojai Valley spread, where he keeps a herd of huge African sulcata tortoises he's rescued, along with other intriguing chelonians.

Friend suits his name; he's an engaging fellow with a mane of white hair tied back in a ponytail and a beard to match it. I wonder why the tortoises—which are state property—go to foster homes. Why not just take them back to the Mojave?

"Once it's in contact with humans it cannot go back in the wild," he is convinced, "because it will carry back to the wild any of the bacteria, pathogens, and viruses it's picked up while it has been in human contact."

Dave Friend dates the trend to bring desert tortoises home back to the 1950s, when Americans hit Route 66 en masse for family road trips. "People would go out into the desert and they would see these things walking around. They'd pick 'em up and take 'em home." Experts like Friend suspect that there are more desert tortoises in captivity than there are left in the wild. Since 2007, the California Turtle and Tortoise Club has registered over five thousand of the animals in private hands, and Friend knows many more hang out in backyards without papers.

I ask Friend the same question I asked Lily: What's the attraction?

"They don't sit up. They don't do tricks. They don't lick your face. They're not warm and fuzzy. They're not a pet. They're a cold-blooded reptile," he says. Clearly, he's made this speech before, and he builds the litany to a tortoise-loving crescendo. "When I look at them I think they may have been here long before the dinosaurs, and they have survived in spite of every-thing Mother Nature has thrown at them. Their largest threat is us!"

At the Crossroads Cafe in Joshua Tree, I order the tofu scram-ble with the Soyrizo-laced potatoes and tortillas to fuel me on the drive across the Mojave. By the time I get to the Ivanpah Valley I've been driving at excessive speeds on the empty two-lane blacktop for about three hours, past the wavelike Kelso sand dunes and the towering piles of the Granite Mountains glowing golden in the desert sun, through endless sage and the Joshua Tree Forest north of Cima. Indicators of so-called civili-zation have been few. The sign for Roy's Motel and Cafe towers over the few miles of Route 66 I hit at Amboy—the gas about twice the price I paid in Twentynine Palms. At Interstate 40 I slow down when I pass under the freeway figuring the Califor-nia Highway Patrol may be hiding out at the on-ramps looking for speeders. A lonely sign flashing an amber light warns, WATCH FOR TORTOISE. It's easy to imagine the vast Mojave expanse as plenty empty and a few acres of solar farm up the road as no big deal.

That opinion changes in a flash when I near I-15 and see three high-rise boiler towers standing like gigantic watchtowers look-ing out over an enormous lake of mirrors planted on the north side of the freeway. There is a road marked PUBLIC ACCESS that bisects the project, allowing for an up-close look. A guard waves me past a checkpoint and cautions me not to pass the sign that marks the beginning of the private road; it belongs to Bechtel, the company in charge of construction. "The rangers will write you a $250 ticket!" Workers in hard hats are scurrying about in road graders, pickup trucks, and tractors—setting up lights for night work and hauling trailers laden with mirrors ready to cover over the former tortoise homeland. Loads of glistening mirrors pass in front of me reflecting the deep blue desert sky, punctu-ated with stark white clouds, along with the torn-up desert around me and the still-unmolested land past the perimeter of the project—land that stretches out to the mountains on the west side of the valley.

"Stay Focused to Reach Our Goal of Zero Incidents," a Bechtel sign screams out to me. I turn my rented Toyota around and head off the maimed BLM land and back across the Mojave, thinking about the contradictions that face us all. How much gas did I use to witness the site for myself and snap a few photographs? How about the carbon footprint of my airplane trip to Southern California? The farm-raised grilled Atlantic salmon that I ate for dinner in Pioneertown at the cheery restaurant Pappy & Harriet's? What did it take to transport it from where— Scotland, maybe—out to the desert? Not to mention the millions of dollars spent to move those seventy-five desert tortoises from the construction site, a move considered a death sentence by the very biologists hired to do the job.

Zero incidents? Sorry, Bechtel and BrightSource. And sorry for the tortoises and the rest of us. It's too late for that zero incidents wish.

• Fred Relieves Himself •

For his first day in Oregon Fred sits in his house on his heating pad. Just sits. When I come over to check on him, he expresses no interest. When I make a sudden move or too much noise, he sometimes pulls his head back toward his shell, but only for a moment. He ignores the raspberry I offer him and acts uninterested in his water dish. So I decide to bother my turtle purveyor one more time. I write Matt Frankel and tell him how Fred was skittering all over the living room before I put him in his box and now in his box he just sits. "Does that mean he was anxious running around the living room," I ask, "and that he's happy to hang out in his house?"

"Not sure," is Frankel's honest reply. "There is some security in limited vision. In the living room he may have been seeking cover. Just let him do what he does and with time you will learn his personality."

I decide to liberate Fred and check if I can discern whether he likes being a free-range turtle or prefers to be cooped up in his house.

I pick him up and his legs start moving frantically as if he is swimming. (Later I learn that's not an appropriate method for lifting a turtle; rather, those legs should have the palm of the other hand available to stand on for a sense of security.) When Fred is back on the living room floor he again moves fast—and heads directly under a chair, where he sits, motionless. Does he like it in the dark under the chair, or is he scared? I put him out in the open again, and he heads for a bookcase, where he noses himself against the bottom shelf, a shelf of cookbooks (none with recipes calling for turtle meat). I figure Frankel is correct, and I pick poor Fred up again, take him back to his house, and report the news to what may be an exasperated Frankel. "Note the words you use for a turtle," he writes. "Wishes. Happy. Feels. Anxious. Free." He offers an alternative litany. "Eat. Poop. Breed. Hide. Bask. Sleep. Not necessarily in order of importance."

But later in the day Fred makes a few moves. He pushes his water dish while exploring the inside perimeter of his house. He stretches his neck toward his basking light. It's a long neck. And he walks (crawls?) toward the saucer Sheila piled with lettuce, placing a foreleg on the dish and resting his head on it. And Frankel's forecast is correct: Fred pooped.

— *Turtle on the Menu* —

The fast-talking waiter's hair is slicked back, his polo shirt and slacks are skin-tight and immaculate. I'm in a *paladar*—a privately owned restaurant—in Miramar, west of touristic Old Havana, close to the seafront Malecón. Miramar is home to foreign embassies and sprawling villas—some maintained or restored to their opulent prerevolutionary sheen, others not so lucky. But this dinner house is first-world, from the white tablecloths to the prices.

"Listen to me!" the waiter orders, equally proficient in English and Spanish when talking food. His take-charge attitude is stereotypically macho. At a time when grocery store shelves lack even basic staples, he reels off a list of what his kitchen offers: beef and lamb, chicken and turkey, pork and boar. I'm listening for the fish and muse about Hemingway when I hear marlin. But as he reels off the main courses I think I hear him say, "Turtle."

"Tortuga?" I ask. "Did you say *tortuga?*"

"Yes," he smiles, "*tortuga.*"

We order. I choose the red snapper. It's seared just right, served with fresh lime and grilled vegetables. "Hotel California" and "Beat It" play softly (if those tunes can be soft even at low volume) in the background. The sweet salt air smell of the Caribbean, just a few blocks north, drifts onto the patio—nice contrasts to the exhaust stink from old Chevys that wafts into my Habana Vieja apartment all day.

While my dinner companions talk and finish their meals I excuse myself and near the kitchen door grab the waiter.

"I'm curious about the turtle you offered," I say, playing a role that is not fraudulent: that of the *Yanquí* tourist. "What type of turtle is it?"

He's happy to talk. "We call it *kawauma* [*sic*]," he reports. I ask him to repeat it and to write the word in my notebook. He's happy not just to spell it but, unprompted, to provide details.

"It's illegal," he says with a conspiratorial smile, "but we always have it."

"Where do you get it?" is my natural next question, and he's happy to answer it.

"We buy it from private fishermen." The fishermen, he explains, shoot the turtles. "The turtles are in the ocean and the ocean cannot be controlled." He's smiling. Cuba's struggling government can't keep roads in decent repair or government-owned markets stocked. It's not even remotely equipped to patrol the high seas, chasing turtle poachers.

"It's the black market," the waiter shrugs off the obvious.

"That's how it works here. The *kawauma* is delicious—just like beef," he gushes, as grilling meat aromas drift toward us from the kitchen. "You must try it. We always have it here. We call it *kawauma* but," he emphasizes the point to make sure I know he's offering the real thing even though he cannot name the species, "it *is* turtle."

I thank him for the tutorial and, in Cuban style, offer my hand. He shakes it. We exchange names. And then it's his turn for a question.

"Are you a reporter?"

I smile and say, "I'm a university professor."

I head back to my table and he to his duties, but not before he demands again with that conspiratorial smile, "Don't use my name."

Later I learn that any fine that might be slapped on a fisherman illegally catching a turtle costs him less than what he makes selling a turtle's meat on the black market. The same calculus holds for the private restaurants. Their fine—in the unlikely event they're punished—amounts to less than the profit on the meal.[5] When my landlords stop by my apartment in Habana Vieja to check on the broken microwave oven that weekend, I ask the couple about *kawauma*. "Critically endangered and protected by law," they exclaim almost in unison and without hesitation. "In danger of extinction and illegal to catch and eat." And they nod solemnly when I ask if there are Cubans who nonetheless eat *kawauma*.

My waiter can't spell. On his menu was *caguama*, the loggerhead sea turtle—scientific name *Caretta caretta*—and endangered. An El Salvador brewery peddled its Caguama brand beer in the North American marketplace and its branding language romanticizes the turtle as a catch. "Legend has it that the fishermen of Central America sought the Great Loggerhead Turtle in warm tropical waters," reads the copy. Legend? How about ancient and contemporary history. "It was tribal belief that this powerful turtle, also known as the 'Caguama,' symbolized good fortune for the fisherman's village." And then the pitch. "It is our hope that you too will experience the good fortune of the Caguama when you experience this award-winning Latin beer."[6] Despite the name and the image of a turtle swimming across the six-pack box, not a word about the loggerhead's endangered status.

— *Palo Turtle Sacrifices in Cuba* —

On another Havana evening, my fixer escorts me a few blocks through the dark and dilapidated streets of Centro Habana to a crumbling apartment house, a typical example both of the Detroit-like collapse of Havana's infrastructure and of the city's overcrowded housing stock. We walk through an unlit passageway and stop at a doorway. My guide says a few words to a woman behind a curtain. The curtain is pulled back to reveal a tiny living room replete with basic kitchen equipment and dominated by what appears to my eyes to be a chaotic mishmash of random items—many impossible for me to identify as I take a first look—and all assembled into what looks like an altar. The air is stifling, hot, and dank. Wet with sweat, I'm invited to sit on a worn café-style wire chair. After just a brief wait, I rise to greet the man coming down a narrow flight of stairs into the room.

Abel Hierrezuelo is a *tata nganga*—a priest in the Afro-Cuban religion Palo, a sister practice to Santeria. Known in the Palo community by his nickname, Muñeco, the priest takes a seat opposite me and next to the altar, smiles, and, based on the credibility of a mutual acquaintance who referred me to him, agrees to explain the role of the turtle in Palo practice. Muñeco is wearing a blue and white gingham cap and a white sleeveless T-shirt—a so-called muscle shirt—and he has the biceps to go with it. A rosary and other necklaces dangle on his chest, beaded bracelets surround a wrist.

"*Jicotea*," he tells me, "is used as food for *resguardo*." It is obvious with his first words that I'm going to need help with terminology and context. Muñeco patiently explains that a *jicotea* is a small turtle that is used in Palo ceremonies—initially while alive. A *resguardo* is an object, he says. It is a "thing" to take out all the bad energy from a sufferer's body. And the *resguardo* is represented by the various items called the *prenda*, or altar—that eclectic collection assembled next to Muñeco. "The *prenda* is used to take all the bad energy out of the house and body." Before Muñeco details the bloody and fatal role of the turtle in the ceremony, he identifies another of the crucial players in the ritual. Siete Rayos—Seven Rays—is the Palo equivalent of Santeria's Changó.

My note taking is slow. I'm checking the spelling of words and names with Muñeco. I apologize for the laborious note taking. "Knowledge is sharing," he dismisses my concern with the reassuring voice of a secure sage.

"The food for the *prenda* is the turtle," Muñeco reiterates. "The only parts used are the blood and the corona." He describes the corona as the bone connecting the turtle's head and the rest of its body. The turtle blood is mixed with dry wine—red or white, it doesn't make a difference. What matters is that the wine is dry, not sweet. I look over at the *prenda*, identifying some of the potpourri on display: a partially smoked cigar, a bouquet of flowers that look like red-orange and maroon daisies, a white hamburger bun like the ones I buy down the street from my Havana apartment after waiting in line at the government bakery, several dolls wearing local garb and adorned with beads, a few walking sticks topped with hats (one of them a shiny purple satin that's decorated with purple and pink glass baubles), and a wooden staff with a hand-carved skull for a grip. "The blood of the turtle is sprayed on the *prenda*," says Muñeco. Turtle sacrifice in Palo is reserved for treating men suffering impotence or those enduring stomach problems or for women with sex and gut disorders.

"Why the turtle?" I ask Muñeco.

"Because it is a strong animal with four legs and good energy," he says.

"And the importance of the four legs?" It's the second time he's mentioned them as ceremonial attributes.

"Four legs are stronger than two," Muñeco explains, without suggesting I should have figured out the obvious. "Four legs are stronger than a rooster, with only two legs, for example."

Muñeco says he's always on call and that ailing clients—all are Palo initiates—knock on his door whenever they seek treatment. "I'm always consulting. When I'm eating. When I'm watching TV. When I'm making love with my wife. People come ask for help and I can't say no. It's my job to help people." He checks them out and determines if they need a ceremony. If so, he organizes it for later that same day. His last turtle sacrifice was the week before. He obtains the turtles from peddlers, who supply them for about 60 pesos—approximately three U.S. dollars. The ceremony starts with Muñeco singing a song to prepare the patient's stomach or sex organs to be healthy. He chants a few bars in a singsong baritone example for me, the words repeating the announcement, "The blood is going into the *prenda*. The blood is going into the *prenda*."

As the *tata nganga* sings these words and prays, the sacrificial turtle is dispatched—usually with a swift and practiced slice of a knife after the priest asks permission of the Supreme Being to sacrifice the animal. This permission is needed in order for the priest to avoid being punished by the oricha that owns the animal—in this case, Siete Rayos, who owns turtles. Some practitioners slit the neck of the turtle to drain the blood; others decapitate the sacrificial chelonian. The turtle remains are placed at the shrine as a physical offering to Siete Rayos.

Skill with the knife is mandatory in order to minimize the animal's pain and suffering. Sometimes—rarely, I'm told, and when it occurs it's usually at rural ceremonies, not those in

Havana and other urban sites—when the priest is possessed by spiritual energy, the priest bites off the turtle's head and spits it out.

The priest sings and prays. Blood from the sacrificed turtle is mixed with 101 powders obtained from the assorted iconic items that constitute the *prenda*: onion and garlic along with other herbs and roots. The blood and powders are mixed with the wine, and the concoction—known as *chamba*—is both swallowed by the priest and petitioner and sprayed from their mouths onto the *prenda*.

Muñeco reaches over toward the *prenda* and pulls out an old vodka bottle filled with a brown liquid. He tips it to his mouth, takes a swig and swallows, then takes another and spritzes it onto the *prenda*. He proffers the bottle to me. I thank him and politely decline. I'm the cautious *Yanquí* journalist who drinks only bottled water while in Cuba, rejects ice in my rum and ersatz-Coke drinks, assiduously peels or cooks my fruits and vegetables, and rarely eats from street vendors. Later I confirm that since I am not a Palo initiate, declining the *chamba* was not taken by Muñeco as an insult to his hospitality.

"Sometimes powdered human bones are added to the *chamba*," says Muñeco, who adds with solemn reassurance, "human bone *chamba* is not designed to be swallowed."

With the dead turtle on the *prenda* and its blood mixed with the wine and powders, the priest draws an ideogram on the floor of the altar room. Muñeco finds an appropriate ideogram for each ceremony he performs from a "book of knowledge," information handed down generation to generation by oral tradition supported by his hand-copied reference book. He goes back up the stairs, returns with a collection of pages, and flips it open to a sample page. Various curved and straight lines—many ending with arrow points—intersect with each other and with symbols that look similar to dollar signs. Crosses and circles are spotted about on the page. It's all drawn in blue ink and the identifier "7 Rayo" is written on the bottom of the page. This is a Siete

Rayos ideogram and is an example of what he draws with eggshell-powder chalk during the ceremony after he uses a conch shell as an oracle to choose the appropriate ideogram for the specific ceremony in progress.

Once the turtle is sacrificed, the same procedure is followed with a chicken, "because the blood of the turtle is hot and strong so a chicken is used to refresh the *prenda*." Next, to conclude the service, all of the participants at the event are cleaned by herbs shaken at them by the *tata nganga*, the priest.

And what happens to the turtle corpse?

It resides for a specific number of days on the *prenda*, a time period set by the priest. The *tata nganga* asks an oracle where the final resting place should be. A usual locale is at the foot of a palm tree or a silk cotton tree. But eating the sacrificed turtle is forbidden because the turtle is sacred. Keeping live turtles in homes is common among believers because turtles are, Muñeco tells me, strong animals that keep secrets inside their shells. House turtles live in plastic laundry tubs or are left free to wander around the living quarters. Water in a house turtle's tub is collected and used to cleanse the home. We're both sweating in the steamy Havana humidity, the air in his cloistered altar room is as still as the ceremonial turtle shell he's showing to me. He continues to suffer my questioning. "I am a son of Changó and Siete Rayos," he says, merging Santeria and Palo. "Sacrificing turtles saves people because animals are complements to human beings."

Palo and Santeria are religion jazz. Neither is confined by a dogmatic orthodoxy certified by an authority. There is no official liturgy. Rituals evolve as they pass from one believer to another. Lessons from Muñeco and others who practice and study Palo and Santeria make it clear that there are at least as many variants to the overarching common traditions as there are mouse-sized cockroaches in my Habana Vieja apartment kitchen. In that spirit I ask Muñeco if there is a negative aspect to taking the lives of turtles. His no is unequivocal. "The turtle," he explains with gracious patience, "takes on the bad energy. The spirit of the turtle—if it's finished with its tasks on Earth—joins the Supreme Being."

More names are added to the playbook. In Palo that Supreme Being is Zambi; in Santeria it's Olofin. "If it's not finished with its tasks on Earth, the turtle is reincarnated into another animate or inanimate being."

Palo priest Muñeco—Tata Nganga Abel Hierrezuelo—and I rise to say our good-byes. We embrace despite the suffocating heat and resultant sweat. I thank him for sharing his beliefs, techniques, and experiences. "This is your house," he tells me, and I tell him he always has a house in Oregon.

— *Santeria Priest in Training for Turtle Sacrifices* —

The next night, just a few blocks down the street from Muñeco's place and around a Centro Habana corner, I'm escorted into the second-floor apartment of the Santeria priest who directed me to live with a turtle.

I clear the last step of the steep and narrow staircase, and I'm led into a cramped altar room as the priest is completing a ceremony in the room adjacent. Despite being uninitiated into Santeria, I'm left free to watch. Two stark-white chickens lie lifeless on the floor. Candles stuck on the floor burn in the otherwise dark room. The priest gathers the two fowl into a white plastic bag. A woman—the petitioner—walks into the altar room and places paper money in a bowl on a table covered with a white cloth, two plates of sliced white bread, a couple of cigars, and various wooden humanesque beaded totems. She returns to the room where the sacrifice ceremony has taken place. The priest—the *babalawo*—hands her the bag and offers her a litany of instructions. She departs down the steep stairs clutching the bag of chickens. He extinguishes the candles, takes a bottle and pours liquid from it onto the floor—onto white markings made with a powder. He chants, washes his hands with the liquid, continues to chant, and next takes a pull from the bottle and sprays its contents from his mouth onto the powder. Then he takes a cloth and wipes the floor clean. Without further ado, the *babalawo*, Sergio Martinez, comes into the altar room and flops himself down on the floor opposite me.

Dressed as casually as he sits and talks, Martinez is wearing a purple T-shirt and burlap shorts. His hat, made of flamboyant golden brocade, is a puffed-out shape reminiscent of a chef's toque, its bright green band decorated with gold sequins. A strand of white beads hangs around his neck. As did Muñeco about Palo practices, Martinez refers to the turtles used in Santeria collectively as *jicotea*. "We use *jicotea* because it is an animal that lasts long and symbolizes time. We use it," he says about the religion's sacrificial rituals, "to save lives when the saint asks for it." The saint that may call for a turtle is usually Changó, but others in the pantheon may seek a turtle for deeds they perform for followers. Saving lives, he says, is the primary goal of a Santeria turtle sacrifice, but other human challenges may call for the strength of a turtle. Martinez offers "unifying a couple" as an example.

Babalawo Martinez provides a straightforward and, thanks to the lesson from the previous evening, a familiar description of the procedure: cut off the turtle's head, spread its blood (and with Changó involved, the blood must be mixed with dry wine) onto a shell that serves as home for the oricha in question. The believer then drinks the turtle blood and wine mixture. The turtle's blood is spread onto stones, and the stones are placed into some sort of receptacle—such as a gourd—to hold the energy of the oricha, which resides in the stones.

"The *jicotea* is powerful because it is very old," Martinez, despite his youthful face, looks professorial as he places the turtle into Santeria context. "It comes from the era of the dinosaur and was one of the first animals created by God. The shell is hard," an example of the turtle's strength, "and it is the animal preferred by Changó. When one is going to be initiated into Santeria with Changó as your oricha, the first deed you do is with a *ji-cotea*." That may be a sacrifice ceremony. It may be that a turtle is placed into a river—alive. "Sometimes," Babalawo Martinez explains, looking at me with his deep brown eyes, "people need a turtle always walking around your house. The turtle collects the negative energy from the house. When the turtle lives in a tub of water you take the turtle out of the water and spread the water at the entrance to change the negative energy." Some believers, Martinez tells me, wash the entryway with the turtle water themselves, "but to do it correctly, you must use a *babalawo*."

To sacrifice a turtle, a *babalawo* must earn the "power of the knife" from another *babalawo*—his godfather, to use the Santeria term for a mentor. At the current stage of his Santeria career, Sergio Martinez is allowed to sacrifice two-legged animals such as the chickens I saw earlier in the evening. "Before you get the knife," he tells me, "you have to buy animals and learn proper killing techniques: goats, deer, lizards, and *jutía*,"—rats native to the Caribbean that live in trees. "I am going to receive the knife," he says with complete confidence. He figures the promotion will come within a year, when his godfather returns from Miami. A *babalawo* who does not get the knife within the requisite seven-year study period languishes. It's a fate that reminds me of the academy and professors who fail to earn tenure.

My lesson complete, we say our adios, and that's when Babalawo Martinez instructs me to live with a turtle. Careening in the old Olds passenger seat along the Malecón toward my apartment, I consider the *babalawo*'s insistent words. "You must," he said about cohabitating with a turtle, "for the good energy and for your health." Later that night, after a marginal dinner of black beans and white rice at the service-challenged Hanoi Restaurant up Teniente Rey toward the Capitolio from my Habana Vieja flat, a dinner Sheila shared with little conversation because of the relentless, repetitive, monotonic assault of the blasting house band (not all Cuban dinner music is of the melodic Buena

Vista Social Club type), I resolve that once I return to Oregon, high on my list of tasks was the *babalawo*'s assignment to live with a turtle.

Meanwhile, I continue my attempts to understand the voracious human appetite for all things turtle, an appetite that motivates poachers and smugglers.

• Fred Finally Eats •

This morning Fred is staying in his cave, the dark part of his house with no windows and just a doorway to his food, water, and basking lamp. I open the top to check on him and he looks okay. He's ignored his raspberry, his apple pieces, and today's offering, a ripe banana. Fred responds to the menu by sitting in his cave, expressing zero interest. All day he sits in the dark. A couple more times I lift the lid and check on him. He looks up at me—maybe with curiosity?

In the evening I'm preparing a salad for dinner, slicing up a head of Napa cabbage. One cut severs a worm hidden from my view between the leaves until I whacked it. "Give it to Fred!" suggests an enthusiastic Sheila. She's been worried about his lack of appetite. I open up Fred's cave again and place two pieces of cabbage-cum-worm-halves in front of him. But my delivery was too abrupt. For the first time I watch Fred pull his head 100 percent into his shell. It disappears and headless Fred backs off from my hand, the cabbage, and the worm parts. "Sorry, Fred," I murmur and close him back into the dark. The moment reminds me of the sign my father affixed to his photo lab: "This is a darkroom. Don't open the door. It lets out all the dark."

After dinner I check on Fred again. The worm is untouched but he's left his cave and walked over to his heating pad, and I find him with a hunk of banana in his mouth. After the banana feast I note him taking a slurp from his water dish. Fred finally broke his fast. What a relief. The next morning the Napa cabbage worm is gone. Good ole Fred!

THREE

THE VORACIOUS CONSUMERS

— *Turtles in the Soup* —

"I've caught a lot of turtles in my life," Cajun chef John Folse laughs when I call him a turtle expert. Chef Folse is famous for his Papoo's turtle soup. I looked to him to help explain why turtle soup draws diners to his restaurant tables. "Tradition," is his quick response. Born in St. James Parish and as famous for his cookbooks as his cooking, Folse founded a culinary institute at Nicholls State University in Thibodaux, Louisiana, a school offering a degree in Cajun cookery. "The English arrived in Louisiana in the 1700s, and they had a really strong connection with turtles as a seafaring people." He stirs the turtle soup publicity pot and waxes on, "The swamp is our pantry." The bayous' offerings include, of course, turtles. The English made turtle broth and used the turtle meat only for flavoring, tossing it out before the meal. The French added roux, the Spanish added spices, and some hungry soul added turtle meat and yelled out, "Oh my God, taste this!" Turtle meat soup was born.

Wild Louisiana snapping turtles survive (*Chelydra serpentine*) in the bayous and swamps, and Louisiana law allows Folse and his crews to catch as many as they need for his soup. "Hopefully we're getting old tough meat because the legendary Louisiana turtle soup comes from long, slow cooking."

Folse tells me in his languid Cajun lilt that he just saw a snapper about three feet in diameter. "His head was at least as large as mine. That size turtle makes the best of the turtle soups because once it's cleaned and you put that meat with the bone in a stockpot and you cook it for four and a half hours, you create an unbelievable robust stock. Then you debone that turtle and start your turtle soup. My fishermen are all getting the large turtles out of the swamp." He says the Louisiana Department of Wildlife and Fisheries rangers consider the snapping turtle population stable and encourage his crews to harvest the larger older turtles because they're no longer mating, laying eggs. Science and experience, says Folse, are changing old Cajun traditions. "When my dad went out, he wanted a five-pound to six-pound snapping turtle because he thought those were the best, and he said, 'Don't ever kill a large turtle because we want to keep those old turtles alive in the swamps.'" But rangers, claims Folse, teach otherwise. "They come to us and say, 'No, no, no, no! Don't ever kill a small turtle. They're the ones repopulating the swamplands. Take away the old turtles because all they're doing is eating more of the foodstuff in the swamps that the younger turtles really need.'"

"Rubbish," is the response of chelonian scholar Peter Paul van Dijk to the chef's assertion that the big old turtles don't reproduce. "I am not aware of any scientifically sound evidence that older turtles, of any species, become 'reproductively senile' or cease reproduction."

Which makes me wonder—regardless of his turtle biology knowledge—if Folse just feeds his customers tough old turtle. Yes, indeed, Folse confirms with robust enthusiasm, he seeks old tough turtles for his recipes. "The best turtle soups and the best turtle fricassees and the best turtle sauce piquante"—all dishes legendary in Louisiana—"come from the long, slow cooking." And this is flesh-in-shell cookery. "The larger turtles are the best, including the shells. Right down the center of a Louisiana snapper, right down the center of that shell, is what

I call the tenderloin of meat inside of that ridge of the turtle's back." Once the shell is separated from the flesh, there's more hand work to be done. "My father and grandfather showed us as young children—young cooks—how to take steel wool and wash the outside of the shell and really scrub that shell down good to get the algae off and then cut that center part of the shell—that ridge of the shell—and put it into the pot with the soup because that meat was the best and the shell would give a little gelatinous texture to the finished dish." Chef Folse says the cast-iron pots of his father and grandfather were the tools of his culinary education.

The cooking lessons came from his family, but Chef Folse learned from his indigenous neighbors how to catch the ingredients. "We used to fish turtle with our feet, as the Indians did. We would go into the little murky bayous," and his soft, slow Louisiana accent conjures images of swampland. "We would go into these little muddy small bayous, and we would walk the banks with our bare feet as we had been taught, looking for those little air bubbles coming up where a turtle would just have its nose coming out of the top of the water. The bubbles told us where the head was. We would walk up and feel the ridge of that turtle's back, and with our foot push the turtle down, and then reach down to get it from each side while it's still under water." It's a technique he still uses when he brings students to the bayous to learn how to harvest turtle. And he teaches his students the Native American regional mythology: if you eat the long-living turtle you're guaranteed a long life.

Celebratory meals at Folse's own family table featured turtle. "When company was coming and we had turtle, it meant that somebody really special was coming. Daddy would go out into the swamp to get the right-size turtle." The good china graced the table for the company and for the turtle dish, and turtle was the talk of the dinner party "because we could tell the story of how we caught the turtle." Folse's enthusiasm for those family turtle days in old Louisiana is boundless. "When we thought of the good times, we thought about turtle. When we thought about company coming, we thought about turtle."

No question Folse was paying attention at those family gatherings, especially to what was happening in the kitchen. "You were judged as a cook by how you could clean and cook a turtle. Nothing is more important in Louisiana cookery than the tradition of handing down the black pot from the best cook in the family to the next best cook in the family. To be the cook of the hunting camp is better than being elected governor in Louisiana."

"It's *especially* better to be a good cook than to be elected governor in Louisiana these days," I say, interrupting his rhapsody. "Consider some of the inheritors of Huey Long's office."

Chef Folse's laugh is hearty. A prime example: four-term governor Edwin Edwards, who was sent to federal prison after the guilty verdict in his racketeering trial. He was notorious for announcing, during his successful 1983 campaign, "The only way I can lose this election is if I'm caught in bed with either a dead girl or a live boy."[1]

Papoo's turtle soup is named after Clarence Bouchereau—called Papoo by his grandchildren. Most of Donaldsonville, Louisiana, knew his famous concoction, says Folse, a recipe that takes four hours to make and calls for two pounds of Louisiana snapping turtle meat to simmer for over two hours (or until tender) and then cook another half hour once all the vegetables, herbs, and spices are added. "Ladle into warm soup bowls and serve hot with toasted French bread," instructs the chef.[2]

— *Turtle as Viagra* —

Cajun cooking is not the only cuisine to feature turtle as a prime ingredient. Worldwide, turtle meat is favored not only for its taste but also for its supposed ability to restore vigor and promote longevity. Turtle meat, shells, and eggs figure prominently in recipes for aphrodisiacs. "They're delicious," a Costa Rican bus driver says as he quaffs a couple of endangered olive ridley sea turtle eggs. He uses the Spanish word for eggs—*huevos*—a word that means the scrambled type and is the vernacular for a form of macho. "They give an active man more potency. They make him more erect! I could drink them all night."[3]

What he doesn't know is that turtle egg salsa could be spiked with Viagra. If such a potion proves effective, a customer— ignorant of the real secret ingredient—understandably credits the turtle, increasing demand for more aphrodisiac made with turtle eggs. Advertising campaigns combat this mythology. The Argentinian actress Dorismar appears in one such an ad sponsored by Wildcoast and supported by a collection of turtle-protecting NGOs. Striking a sultry, come-hither pose in a skimpy swimsuit, she purrs: "*Mi hombre no necesita huevos de tortuga marina.*" "My man does not need sea turtle eggs," is her message, "because he knows they don't make him more potent."[4]

But belief that turtle eggs and turtle meat heighten sexual performance and satisfaction is not limited to provincial laborers. When a distinguished academic at my university learned that I was researching chelonians and human behavior, he leaned close to me across the conference table during a meeting of just the two of us in his private office and—sotto voce— confided that he had eaten turtle and that there was no question in his mind that credit went to the reptile for the extraordinary sex play he enjoyed after the meal. Pity the turtles for their reputation.

— *Voracious Collecting* —

From gourmet menu items to exotic medical treatments, from companions to haute couture, turtles are in great demand, creating markets for turtle hunters—both law-abiding traders and scofflaw poachers. In modern-day China, for example, the turtle plays a pivotal role. The growing and affluent upper middle classes seek turtles as status symbols. Manic capitalism in "communist" China (and all the good and bad that comes with it) is compatible with ancient turtle lore and collecting. Masses of turtle fans there are awash with cash. "Chinese culture is bad for turtles," is how Ross Kiester, chief scientist at the Turtle Conservancy, summarizes his concern in an era when China is an easy-to-target bogeyman. It's Kiester who insists "Turtles are becoming the orchids of the animal kingdom." Over bánh mì sandwiches in an Oregon Vietnamese eatery, Kiester is regaling me with stories of his fieldwork in Asia. Most hybrid animals—like mules, the issue of horse and donkey—cannot reproduce. Not so with turtles. Breeders can mate a hybrid—a turtle with parents of two different genera—with a another turtle of a third genus, creating hatchlings that are trigeneric. "So, as with orchids," Kiester explains, "you get designer turtles—thousands of different types of turtles." That biological fact may be good for the turtle-peddling business, but it is not without risks. Mother Nature may add her own unintended consequences to the mix.

What motivates maniacal collecting, legal or otherwise? Stephen Calloway, a curator at the Victoria and Albert Museum in London, who himself collects correspondence from the 1890s, says, "Collecting is for most collectors just an absorbing and largely harmless pastime. But for a few," he cautions, "it can become a dangerous preoccupation. Its sensations—the breathless exhilaration of the quest, the thrill of capture, the enjoyment of novelty, the sense of satisfaction and pride in possession—are as addictive as any drug."[5]

Laws and regulations in the United States restricting the varieties of turtles that can legally be hunted for food and trade can be stringent. The demand, coupled with the scarcity of turtles legal to trade, makes them ever more valuable to smugglers.

Underground marketplaces thrive for those willing to break the law to sell and to buy precious turtles.

Turtle references in my previous books about human interactions with other animals caught the attention of Ross Kiester and Eric Goode, founder of the Turtle Conservancy. I met with them when I was on a trip to New York City, and after encountering their infectious enthusiasm, I started paying more attention to turtles and tortoises.

Eric Goode brings panache and glamour to his turtle research and rescue career from his days running the celebrity-packed New York City nightspot Area and swanky downtown hotels like the Bowery. The Conservancy publishes the ultra-slick journal *The Tortoise* (I serve as a contributing editor), throws the star-studded annual Turtle Ball as a fund-raiser, and operates the Behler Chelonian Center in Southern California. Dr. Paul Gibbons, who is managing director and principal veterinarian of the sprawling turtle refuge, oversees the captive breeding and care of an enormous variety of rare turtles, many threatened and endangered. Gibbons also heads up the Plowshares Tortoise Species Survival Plan, a multiagency initiative of the Association of Zoos and Aquariums. The ploughshare or angonoka tortoise (*Astrochelys yniphora*) is facing imminent extinction; it is "the world's most endangered tortoise," according to Goode.

Ploughshares—illegal to take from the wild, sell, or buy—fetch huge sums on the black market. Fifty thousand dollars a turtle is an oft-quoted figure for the high-domed tortoise with the distinctive ploughlike protrusion from its plastron jutting up under its chin. Their remote Madagascar homeland is protected by local rangers who are inadequately trained, equipped, and paid. Imagine a tortoise worth $50,000 in a country where the median annual household income is about $1,000.[6] In an effort to combat poaching and smuggling, Gibbons and his colleagues at the Durrell Wildlife Conservation Trust began defacing ploughshares in 2012, using a Dremel drill with a rotating burr tip to engrave letters and numbers into the shells of at-risk tortoises. "The carving is deep," Gibbons explains, "more than a quarter of an inch deep on an adult, down to where the keratin becomes soft. It's very hard on the surface of a healthy normal tortoise, and as you get closer the bone starts to get soft."

The marks are both disfiguring and identifying but the cuts do not hurt or harm the animals. The hoped-for result was that poachers would avoid the marked tortoises. If they still stole them from their reserve, it was hoped (and expected) that the marred animals would be worthless to buyers, especially in Asia. Gibbons and his coconspirators theorized the markings would be interpreted as a code that decreased the tortoises' "spiritual energy." The code also identified the animal and made it possible to track illicit sales. Not long before we talked, a local chieftain was detained for trying to sell an engraved plough-share. Another bad actor strolled into a Madagascar veterinarian's office with a snatched ploughshare and asked, "Can you guys remove this engraving?" A flabbergasted staffer blurted out, "That's illegal," and the hustler grabbed the tortoise in question and ran. A slow learner, he showed up at another vet's with the same request. By this time word had spread. The second office stalled him and called the police. He was arrested holding the tortoise.

Even if the ploughshare rustler had managed to erase the letters and numbers he might well have been caught and the tortoise found. Most marked ploughshares are injected with a PIT tag—a passive integrated transponder. PIT-tagged plough-shares can be tracked and traced with a receiver that captures the radio signals the PIT-injected tortoise transmits.

Gibbons, who is Harrison Ford handsome and looks like he just stepped out of *Raiders of the Lost Ark*, grew up in southern Illinois catching snakes and frogs and turtles while dreaming of making animal care his career. "I was interested in being the James Herriot kind of veterinarian, driving around the country, taking care of people's animals, having all these wonderful stories about the characters that I met and the cases that I had.

That motivated me." It was the people, not the snakes, frogs, and turtles, that led him to specialize in reptiles. "I'm not sure I was any more passionate about reptiles than I was about birds or exotic cats or anything else. But I really liked the reptile people. Reptile people are just loving. Loving. Welcoming. Just wonderful people."

Wonderful people and wonderful chelonians. As Gibbons and I talk turtles he's quick to list the turtle attributes that make those particular reptiles intriguing for him. Most turtles are friendly. They exhibit neoteny: "They look like little babies. They have big eyes and they're kind of helpless." Yes, some turtles are aggressive and bite, but "as a rule they're sort of helpless." Plus, swoons Dr. Gibbons, they're beautiful, showing off "lots of different colors and shapes." Not that he gets emotionally attached to individuals among the thousand or so in his flock. He maintains an academic distance from each, even as he watches them exhibit what he's convinced are distinct behavior patterns and personality traits.

"They like to solve problems," he says.

"Turtles solve problems?" I'm intrigued and seek an example. It's right in front of us.

"We have our Galápagos tortoises," he gestures toward the grazing giants in their verdant enclosure across a wandering path from the picnic table where we're talking, "and we sometimes hang opuntia pads for them. A tortoise would much rather reach up to try and eat a cactus than eat a pile of greens on the ground. It's just more interesting. There's grass all around them and they can eat the grass if they want, but they'll work and work and work and reach up trying to eat that opuntia."

"Are they thinking through this quest?" I ask, "Are they plotting?"

"It's impossible to know what's going on," he acknowledges, scientist that he is, "but I think, just as a rule, animals would prefer to do something interesting than to do the thing that's boring."

Not too long after we contemplated boredom avoidance, I pull off the road to watch a hawk that caught my eye. An off-shore wind is steady and strong over Bodega Head, where I just took a hike. As is usually the case, soaring and gliding hawks accompanied me. But now, prominent in the bright winter sunshine, the lone hawk hovers about a hundred feet up—wings spread, taking advantage of that headwind, not flapping those wings but moving them just enough to control its hover in the gusts. What is it doing up there? Stalking or playing? Both, perhaps, and, at least from my perspective, clearly not bored.

— *World-Class Smuggling* —

Anson Wong was waiting for his bags. Wong is a notorious wild-life smuggler who specializes in reptiles. His luggage popped open in Kuala Lumpur as it was being transferred to another plane bound for Jakarta. Out crawled ninety-five boa constric-tors and the baggage handlers yelled for the cops. Imagine the screaming panic.

Instead of collecting his baggage, Wong was hustled off to court, convicted of trading in the endangered species without a permit, and sentenced to six months in jail—a proverbial slap on the wrist for this international trafficker. What occurred next was a surprise to most seasoned observers of the Malaysian wild-life market, which is famous for its all but wide-open illegal animal trade. This time, prosecutors appealed Wong's light sentence, and the appeals court agreed to change the penalty to five years in prison.

The rationale for the five-year term was not based on worry about the health of the boa constrictor population. Justice Mohtarudin Baki of the Shah Alam High Court ruled that the lower court neglected to take into account the great number of snakes packed into Wong's luggage. Ninety-five boas is one big mess of constrictors. Justice Baki ruled that Wong's haul posed a danger to aviation. Passengers, crew, and airport staff, he opined, would have been in danger if the snakes escaped unde- tected. Life imitates art, of course. Remember the movie *Snakes on a Plane*? Remember Samuel L. Jackson's famous line? "Enough is enough! I have had it with these motherfucking snakes on this motherfucking plane. Everybody strap in. I'm about to open some fucking windows."

But it not was just the "motherfucking snakes" in Wong's lug- gage. Packed in with the boas (and two rhinoceros vipers) was a fat-headed, flat-headed freshwater matamata turtle. With its beady eyes and long nose (to get air so it can breathe while it hides the rest of itself underwater), the matamata (*Chelus fim- briata*) looks to the untrained eye more like pond scum than a prized possession. It may not be cute, but turtle fanciers are in- trigued by its feeding habits. They like to watch their matama- tas suck whole fish into their big mouths, eject the water that came with the fish, and swallow the fish. The matamata does not chew its food well; in fact, it doesn't chew it at all. There's money to be made in the matamata trade, either importing them from their South American habitat or breeding them in captivity. They're sought after as long-living, weird-looking, pretty big (a couple of feet long) exotic pets. Perhaps the matamata travel- ing with Anson Wong was a bonus animal for whomever was getting the boas. It is not endangered and easily available for a couple of hundred dollars.

I had never heard of the matamata before I started focusing on turtles. Funny how once we stumble into a subculture we no- tice things everywhere that were invisible to us before. My best example is the old Studebaker I bought as a lark (joke intended) one day. Suddenly I saw those two-headed Raymond Loewy de- signs on streets and highways wherever I drove. The phenome- non is called "frequency illusion" by Stanford University linguist Arnold Zwicky: "Noticing things that are salient to us . . . looking for things that support our hypotheses."[7] Outside of the acad- emy it was dubbed the Baader-Meinhof phenomenon by a reader who wrote to the Twin Cities *Pioneer Press*, explaining that he talked to a friend about the German gang and the next day his friend read a newspaper story that mentioned them (long after they had faded from the news).[8] Now I see turtles everywhere.

While Anson Wong was in prison I worked at conducting a jailhouse interview with him. But while I was still negotiating with Wong's wife and studying workarounds through the prison visitation rules, the smuggler walked out of prison a free man. An appeals court reduced his sentence to time served—almost a year and a half. Activists struggling to curb the infamous illicit trafficking in endangered species out of Malaysia were crushed by the court's decision. "Mr. Wong and his ilk are not just trading common house pets or livestock," fumed William Schaedla, a former Southeast Asia regional director of TRAFFIC, the wildlife trade monitoring network. "They are dealing largely in species at imminent risk of extinction."[9] Wong's lawyers persuaded the appeals court to spring their client based on their insistence that his bags of boas marked a first-time offense. The first time he was caught in Malaysia, yes. But the court chose to ignore his prior animal smuggling conviction and almost six years of imprisonment in the United States. Schaedla is convinced prosecutors bungled the Wong case—they seized his mobile phone and laptop, but, he says, there is no evidence that they searched them for Wong's suppliers and customers. "The government often complains that Malaysia is portrayed as an illegal wildlife trade hub. If this is the way it handles one of the most high-profile and straightforward wildlife crime cases ever to fall in its lap, it has only itself to blame."[10]

— *Collectaholics* —

Whether the stash is stolen art, butterflies, or turtles, what type of people want to own exotic stuff that's dangerous or illegal or both—especially animals? The lure of the forbidden can be seductive.

"I just wanted these things," Hank Molt told Eric Goode as he tried to explain his lust for rare reptiles, a lust that lured him into years of illicit animal trafficking. "They came from romantic places: Singapore and Indonesia and the Congo." He insisted his motivation as a reptile smuggler was for more than money. "I wanted these animals to have them. And it was an ego thing." His animal expertise was self-taught and, he said, he was disparaged as an amateur by academics and zoo staff: "Everywhere I sought employment in the field I was rebuffed." He opened a pet store and dreamed of stocking it with rare reptiles. "I always wanted what no one else had." Smuggling, he told Goode, became his route to getting not only what others didn't have but also what others wanted—and what he learned they would pay big bucks to buy from him.

Hank Molt sent clients catalogs of what he offered for sale from the animal kingdom. "I was a fanatic about quality," he said, recounting the replacement guarantees he offered clients. If the animal died while the guarantee was in effect, he replaced it. "I would describe something to the toenail," he said about his price lists. "Iridescent labial pits," he offered as an example. "People would say, 'I bought that python because of the way Hank described the labial pits.'" Molt's desire to collect sent him around the world as he stocked his business. But he identified the laws designed to conserve species—like the Endangered Species Act—as a gift. "Without the laws, I wouldn't have had a business. My business was circumventing the laws."

The psychiatrist Werner Muensterberger collected African art. He also studied himself and fellow collectors in researching his book *Collecting: An Unruly Passion*. "They [collectors] like to pose or make a spectacle of their possessions," he believed. The possessions are stand-ins for the collector: "Their deep inner function is to screen off self-doubt and unassimilated memories." According to Muensterberger, collectors suffer from problems of self-confidence and self-esteem.[11]

Obsessive collecting can be benign, of course. During a research trip to Arizona I stayed at a nondescript hotel in Tucson. Sharing it was an impassioned group of (mostly) men who looked as if they came of age in the 1960s. Many wore regalia celebrating their favorite automobile. They're were there for a regional meeting of the National Corvette Restorers Society, showing off their cars and winning blue ribbons for the classiest chassis. Yes, these guys are serious collectors. But their Corvette hobby isn't illegal. It's not jeopardizing the existence of an endangered species. In fact, they're preserving history.

While in Tucson I considered turtle soup for dinner. I had identified a local landmark—O'Shaughnessy's Steakhouse and Piano Bar—where chef and owner Sam O'Shaughnessy added turtle soup to the menu from a recipe he learned working in a kitchen south of the border in Nogales. From a review in the *Arizona Daily Star* I learned it was an affordable $11 a bowl and politically safe: it's made from farm-raised Michigan snappers. Nonetheless, for a fellow who hasn't eaten meat since Jimmy Carter was president, I was not looking forward to dinner at O'Shaughnessy's, even in the line of duty. The restaurant is closed Sundays and Mondays—the two days I was scheduled in Tuscon. I was spared not just tasting the soup but also another of the restaurant's claims to fame: its crooning wait staff singing Sinatra classics.

Yet I must confess it could have been a memorable evening. Candlelight. A fine pinot noir. Farm-fresh turtle soup. And my waitress singing that Cole Porter classic "At Long Last Love," with its apropos signature question for us all: "Is it the good turtle soup or merely the mock?"

Despite my romantic reverie, deplorable conditions for turtles are the norm in tawdry marketplaces worldwide.

• **Fred Versus the Walmart Night Crawler** •

Worms for Fred: I find night crawlers in the Walmart fishing department, Canadian night crawlers packaged by the DMF Bait Company (the *M* in their trademark is a rendering of a worm) in what looks like a yogurt container. "Our worms catch fish or die trying," is their slogan. I fish one out of the carton and place it nose-to-nose with Fred. Fred at first expresses some interest, his head in slo-mo, stalking the worm. But the night crawler acts unfazed and stalks Fred back. It lifts its head cobra style and checks out Fred's face. It does the same with its tail (and it's difficult to determine from a decent distance which is which). Then Fred acts uninterested, turns away from the worm and spends several minutes facing the wall of his house.

Fred looks back toward the worm, walks toward it with purpose, and attacks. He bites at it, holds it down with one of his forelegs, gets it into his mouth. Bite by bite, the still wriggling worm becomes shorter and shorter, until Fred takes a last bite and swallow and the worm disappears. I watch Fred looking around, his head raised and moving side to side, scanning—is he seeking more worm? I realize the drama took place on a piece of newspaper announcing the winners of a local charity raffle. "Congratulations to the winners," reads the headline. Tonight that includes Fred, who takes a foreleg and rubs his face, wiping his mouth.

FOUR

THE TAWDRY MARKETPLACES

— Animal Flea Markets —

"Don't eat before you go to a Guangzhou animal market," I am warned repeatedly by colleagues familiar with the city's open-air marketplaces. "The place will nauseate you. Seeing those animals will make you terribly sad."

They are correct. Asian open-air bazaars are chaos—among the sounds, the smells, and the calamitous conditions for animals in trade, sales pitches of vendors mix with the mournful wails and songs of suffering animals.

Litter after litter of puppies and kittens jammed into tiny cages. Row upon row of boxes filled with mice and guinea pigs, rabbits and ducks. Plastic tubs full of water packed with fish, alongside stacks of aquariums stuffed with larger fish. And, of course, turtles. Bins and bins jammed with all kinds of turtles, just turtles, nothing more, and a little water. Live turtles with barely room to move, turtles piled on top of each other, turtles being handled as if they were produce like apples and oranges— just another commodity. But it's a commodity groping at the slippery sides of the plastic tubs, a commodity relentlessly trying to escape its fate, a commodity lying listless, looking lifeless, awaiting the soup pot or worse.

When my colleague Janet Wasko was in Guangzhou for an

academic conference, she surveyed the turtles on offer at the animal market and was deeply dismayed by the displays she witnessed. Turtles—live turtles—stacked with no apparent regard for their well-being. Yet revered. "Longevity," was what she heard from everyone she asked about the value of turtles. "Turtles have long lives," she was told repeatedly, "and therefore they should be respected." And not just longevity, but also good fortune. Throughout the animal market she found little turtle replicas made of wood or metal perched atop a stack of coins by merchants seeking luck.

Of course it's understandable to link good fortune and longevity with turtles—lucky guys, turtles, they enjoy long lives. Yet none of the Guangzhou turtle sellers or their prospective customers seemed concerned about the sardine-can conditions these good-luck charms were lingering in. "There was no sense that they were doing anything inappropriate or unethical," Wasko tells me when we meet after her trip as she shows me her snapshots of animals and their parts.

From suburban American shopping center emporiums like PetSmart and Petco to mom-and-pop reptile stores to the teeming Asian animal markets to internet dealers peddling turtles, extraordinary amounts of money change hands in the turtle trade, legal and otherwise. But as with the traffic in drugs and weapons, the law does not end illegal business. Mong La, the Tijuana of the lawless northeast region of Myanmar on the Chinese border, is a notorious example of an Asian black marketplace. Everything is for sale: sex, drugs, and animals—alive and in parts—turtles included. But the grotesque occurs in North America, too.

— *Lifelong Turtle Scar* —

This next turtle story promises a hero and a happy ending despite the fact that it begins with the slicing of steaks off live turtles. Such carving to guarantee fresh meat is a butchering practice animal rights activists have been fighting since at least 1997, when a schoolgirl in Vancouver, British Columbia, first exposed the practice.

Just before Christmas eighth-graders from Abbotsford Christian Secondary School headed off to Chinatown on a field trip. The purpose of the outing was to provide students with firsthand experience about Chinese culture, but they saw much more than they and their teacher had anticipated. According to *The Abbotsford News*, when Jayde Crew and her classmates stopped off at a meat market, they were shocked; a live turtle was being sliced into steaks while they watched. No big deal, the butcher told them, routine. "Why don't you kill it first?" asked an incredulous Crew. "It tastes fresher like this," was the butcher's pragmatic response.[1]

Almost twenty years later I caught up with Crew, now Jayde Crew-Nussbaum, at her home in British Columbia on maternity leave with her year-old baby. Despite the intervening years, that day in Chinatown was still a vivid memory. "I remember it very clearly," she recounts. "We came upon this table and this man standing at a table on the sidewalk in front of a store. The turtle's shell was off and I saw him cutting pieces out of the turtle." Pieces of the turtle's flesh. "I saw the head moving and this little turtle face looking at me. I went up to the guy and I said, 'Why are you doing this?' And he said, 'It tastes fresher when the animal is alive.'" Crew-Nussbaum, who calls herself a "huge animal person," was horrified and disturbed by what she witnessed.

Their teacher had equipped students with disposable cameras (in that pre–cell phone era) to document their day in Chinatown. Jayde's mother was an animal rights advocate who had taught her to be an activist, but she could not bring herself to photograph the bloody turtle. "I was extremely upset and I felt like I couldn't look at it anymore," she said. "I remember asking my classmate Joel if he could take the photo for me." Joel snapped a couple of pictures, and when she returned home that afternoon, she shared the ghastly experience with her mother, who grabbed the phone and called authorities. "It wasn't the fact that it was gross" that traumatized Crew-Nussbaum. "It was more the fact that this little turtle was alive and was being cut up. That was my only concern. I remember seeing the gross part, but that's not what resonates with me." What Crew-Nussbaum did not know until she and I talked those years later was that Josh's photo and her story had changed the rules.

"Turtle-Slicing Upsets Kids," was the understated headline in *The Abbotsford News* story about the unexpected field trip lesson, a trip Crew-Nussbaum's mother, Ronwyn, called "pretty traumatic." The paper quoted Christine Schramm, manager of the Rainforest Reptile Refuge Society, a British Columbia shelter for reptiles. "They feel the same emotions we do," said Schramm, including anger, frustration, jealousy, and hatred. "I think something should be done about this." What the schoolchildren witnessed was no anomaly in Canada. A couple of months later, Humane Society inspectors in Toronto, Ontario, discovered a turtle for sale: it was missing two legs. That did not surprise Shelagh MacDonald, program director at the Canadian Federation of Humane Societies: "In seafood markets across the country, live softshell turtles are cut apart and sold as a delicacy," she reported. "Customers frequently request a particular part of the animal. That part is cut off the live animal and the rest of the turtle is tossed back into the tank to die a slow and painful death, or placed in a freezer for future consumption."[2]

At the time of the Abbotsford student outing to Vancouver's Chinatown, turtles were not included in Canadian regulations requiring animals used for food to be humanely slaughtered. What does the government now consider a humane approach to dispatching a turtle destined for the dinner plate? The response from the British Columbia premier's office sounds almost as nasty as the butcher's knife. The turtles should be stunned before being killed, by crushing of the head prior to decapitation. "Ministry staff suggested stunning could be done with pliers."[3] In a notice to turtle importers, the BC Ministry of Environment, Lands and Parks insists that "wildlife should be humanely dispatched," and the word "killed" is added parenthetically to make sure "dispatched" in this context is understood. "A humane death," explains the ministry, "is considered such if instant and painless. It is important to note that turtles are difficult to render unconscious as their brains are resistant to anoxia," which is defined for the importer as a lack of oxygen. "The animals can remain alive for quite a long time even if decapitated. Hence," continues the ministry's instructions, "turtles should be killed [by] stunning, followed by immediate decapitation and crushing of the skull." Not that the ministry advises this as an activity for school field trips. "Dispatch of the Softshell Turtles," concludes the notice, "should always take place out of view of the public."[4]

The publicity and lobbying following the notorious Abbotsford school outing resulted in the Canadian Food Inspection Agency banning importation of live softshell turtles into Canada, the only turtle that authorities believed suffered the live-butchering abuse.[5] This change in regulations was unknown to Crew-Nussbaum until we talked a generation after her encounter with the Chinatown butcher. "Oh, wow!" she said on learning the news. "That's good. I'm glad my efforts did something." A pretty decent legacy for a twelve-year-old schoolgirl. But memories of that day in Chinatown still haunt Crew-Nussbaum. "Yes, definitely," she tells me when I ask her if she still thinks about that turtle looking up at her as it was butchered. When talk turns to animal rights she remembers the dying turtle. "We went to Hawaii and saw turtles and I thought about it then. I just did an SPCA fundraiser walk and I thought about it then. I think about it all the time, really. It's a pretty traumatic thing for a kid. I don't get traumatized when I think about it now, but it still does bother me." She continues to call humane societies when she sees animals mistreated. "I'll never forget that image." The butchering, it turns out, is not Crew-Nussbaum's only turtle story. "I'm from South Africa, and I had a pet tortoise as a kid. It was huge and roamed around the backyard." So the incident in Chinatown "really struck home. He was the family pet."

The shots Josh snapped with the throwaway camera triggered the staff at the Vancouver Humane Society into action. Debra Probert, the current executive director, was an assistant to the founder at that time. She and I meet for a leisurely luncheon at Heirloom Vegetarian, a Vancouver eatery that boasts "a dedication to using thoughtfully chosen fairly traded ingredients with a local and world conscious initiative." No turtle on offer here. "Nobody could believe it," Probert remembers. "It was such a horrible thing to be slicing meat off a live animal. We couldn't understand how that was happening. It was so shocking. They did it while those schoolkids were there, so it must have been in their minds acceptable or they wouldn't have done it. These kids were shocked." The Humane Society attempted to press charges against the butcher and the shop where he worked, but they couldn't. "We tried to find a law that they were breaking and there was no such law. Anti-cruelty legislation at the time was so weak. In Canada animals are property, and a line is drawn between food animals and other animals. This was a food animal." But the publicized spectacle of carving up a living animal in order to satiate the tastes of demanding customers was the catalyst for change. "You just don't cut up animals alive," Probert says. "It's such an unimaginable thing to do." We're finished with our meatless lunch. "Think about it in terms of your own body. If someone were slicing up bits and parts of it—it's just so horrible to contemplate." Given the images our conversation is conjuring, it's lucky lunch is over and not surprising that we don't order dessert.

But before we say good-bye, Probert shares her personal turtle story. "When I was a kid," she begins, and with a nod to the ubiquity of her tale she adds, "I don't know about you—we used to have turtles as pets. We used to go to the store and buy them and I'd feed them raw hamburger and they'd die and I'd get another one." "And you kept them," I interrupt, "in a little plastic bowl." "With a palm tree," she says, finishing my sentence and smiling at the shared memories.

Early the next morning I prowl Vancouver's old Chinatown and find no live turtles offered for sale. At Pender Seafoods, one of the outlets targeted by the Vancouver Humane Society after the Abbotsford field trip, the storefront is shuttered with an iron gate. The window awning is home to pigeons and their feathers and droppings. Faded images of lobsters and crabs (but no turtles) adorn the torn awning. The out-of-business shop windows are covered with sheet metal. On the steps up to the front door, two ancient Vancouver phone books fester and yellow. Other meat markets in the neighborhood are open, but I find no turtle meat—dead or alive. South of Vancouver, in the suburb of Richmond, where about half the population identifies as ethnic Chinese, again I find no wet markets offering turtle. A store called Exotic Meats seems a candidate, but the only things close to exotic are tripe and dead ducks. Of course, to find forbidden turtle I probably would need a password to get into any back rooms that may still butcher live turtles, whether in Richmond, in Vancouver, or in my own hometown Chinatown in San Francisco.

— *A Chinatown Turtle Tour* —

In 2010, when California's Fish and Game Commission decreed it illegal to import live turtles into California for food, some Asian merchants and politicians called the ban discriminatory. State Assemblywoman Fiona Ma said the ruling appeared to disproportionately target Asian American businesses. "These minority markets have had this practice for hundreds of years, and all of a sudden the commission comes up with this policy," she complained in a letter to the commissioners signed by five other Asian American state legislators. The commission—which operated in an advisory role to the Department of Fish and Wildlife—cited the release of non-native turtles into the California landscape as the rationale for the ban, but Assemblywoman Ma interpreted it as an attack on her constituency, especially since grocers stocking live turtles were already required to slaughter them before selling the meat to customers. "I understand the non-native species concern," she wrote, "but they don't ban the importation at pet stores." The Los Angeles Society for the Prevention of Cruelty to Animals applauded the ban, and its president, Madeline Bernstein, joined Ma in targeting pet stores, but she was relentless in her charge that merchants were selling turtle flesh in violation of the law. "The Asian community promised to be good and abide by the rules, and there is case after case of that not happening," charged Bernstein, insisting that "the animals are not fed, not cared for, and the soft turtles are being flayed alive."[6] Flaying turtles alive—even if they are indigenous to California—violates the state's penal code, which orders that "no animal will be dismembered, flayed, cut open, or have its skin, scales, feathers, or shell removed while the animal is still alive." But the penalty for such ghastly marketing barely constitutes a slap: a warning for the first offense "in a written language that is understood by the person receiving the warning"; a fine of up to $1,000 for a second offense, but that cost is waived if the culprit takes a course "administered by a state or local agency on state law and local ordinances relating to live animal markets."[7] Shouldn't that first offense constitute such a lesson?

The next year John McCammam, the director of the Department of Fish and Game, overturned the Fish and Game Commission ruling and ordered that permits again be issued for importing live turtles into California destined for the dinner table. Action for Animals, a one-man animal rights advocacy organization that lobbied for the ban, said the director, who serves at the pleasure of the governor, succumbed to political pressure, "playing the race card." Eric Mills, who founded Action for Animals "because they can't speak for themselves," is not just concerned about the well-being of the turtles but was alarmed by the salmonella, *E. coli*, and giardia found in turtles purchased at Asian American markets in northern California. "I'd sooner eat a dead rat than an animal from the markets," was the nasty picture Mills painted after he saw the results of necropsies that he contracted for with scientists at San Jose State University.[8] When he and I first talked, Mills said he'd seen "cruelty that's unbelievable" in San Francisco's Chinatown markets. "I saw live turtles being cut up with a band saw." It's the same in Oakland, Sacramento, Los Angeles, and San Jose, said Mills. "There's almost no enforcement. It's a brutal, brutal trade and completely unnecessary." As a child Mills was influenced by the high mortality rate of the turtles his parents bought for *his* plastic habitat with the green palm tree. "They all died," he remembers, before returning to his contemporary cause. "They've been here two hundred million years, and we're going to wipe them out for food within a couple of decades." He calls us a dysfunctional species.

It's the first storm of the California rainy season. It's a blustery downpour and my umbrella is broken, so I'm trying to stay dry under the awning of an apartment house at the corner of Webster and Tenth Streets in the heart of Oakland's Chinatown. Just an easy drive or subway ride across the Bay from the glitz of Grant Avenue, the main street in San Francisco's famous Chinatown, this corner is no destination for buses jammed with camera-clicking tourists seeking a glimpse of the exotic. Here the storefronts bear signs for Draline Tong Herbs, the Shan Dong Restaurant, the LG Supermarket. Nothing is designed to lure sightseers as shoppers brace against the rain for daily supplies. Delivery trucks sit double-parked, graffiti competes with bilingual advertising for the Elite Hair Salon, the B&Q Hat Shop, and Liu's Acupuncture and Massage. I'm braving the storm to meet Eric Mills for a field trip through the wet markets that he's been fighting since shortly after he migrated to

Berkeley from his native Kentucky. (Mills likes to tell jokes, including those about his old Southern homeland. An example: Young fellow from Georgia steals a car, drives north, and ends up in New York. He's pulled over by a traffic cop, who says, "All right, kid, you got any ID?" And the kid says, "'Bout what?")

Our first stop is the New Sung Hat Market. I follow Mills into the store, past displays of lovely-looking fruits and vegetables, many varieties unknown to me, their prices and identities marked only in Chinese characters. Mills, with his wire-rimmed spectacles, white beard, and stately stride, triggers no warnings among the sales help as he saunters through the dry goods section to the meat department in the rear of the store. Under a cold case he points to a tank of live frogs. Next to the piles of frogs is another box with a cover on it. Mills doesn't ask for help; he just pushes aside the cover and inside we find live softshell turtles, waiting. On the floor across the aisle we find a couple of plastic laundry tubs jammed with live football-sized turtles. "Red-eared sliders," Mills identifies them. "Wild-caught," he adds with authority, telling me to look at their weathered carapaces as proof that their provenance is not the safety and security of a farm. There is no food in the tubs, and the only sign of water is the damp bottom of the containers. Only one of the turtles is active, climbing on the others, sticking out its neck, searching for options. The others are lethargic, and one looks dead—flopped on a neighbor, its head hanging out of its shell, limp. Mills, though, is animated, expressing his disgust, lamenting the plight of the turtles, railing about their deplorable living conditions, and worrying about the diseases he's convinced they likely carry.

Our next stop is down the street and around a corner, at the bazaar called New Tin's Market. Tarps in front of the store protect bins and bins of lush vegetables and fruits. Once inside, Mills and I head for the meat section, where clerks net live fish out of big tanks for customers' approval. Again the turtles—both softshells and red-eared sliders—are stacked in plastic tubs. There's a drizzle of water in the bottom of the containers, but no food or drink. Several blocks from the heart of China-town we make a final inspection at the Sun Hop Fat 1 market to see more plastic bins filled with more softshells and red-eared sliders, lorded over by a portly poker-faced butcher in a white apron protecting a blue smock, a matching blue baseball cap on his head.

"How much?" I ask Blue Smock about the sliders.

"Twelve dollars."

"How old are they?"

"Dunno," he answers, without changing his dour expression.

"Where do they come from?"

"Dunno."

"You kill 'em here?" I ask.

"No."

"I buy 'em, take 'em home, and kill 'em there?"

"Yeah," he says.

"How?" I ask.

"Dunno," is his final response.

Eric Mills isn't surprised by the exchange. Although offering customers the option of taking live turtles home for last-minute slaughter is against the law, Mills is convinced the practice con-tinues, and a bilingual warning sign in the wet markets suggests he's correct. RELEASING LIVE TURTLES OR FROGS INTO THE WILD, PARKS, OR ANY PUBLIC WATERS IS PROHIBITED BY LAW, it says.

Over a coffee after our expedition to the markets, Mills ru-minates about turtles. I wonder if the turtles mind being piled on top of each other in the Rubbermaid tubs as much as we do. Are they aware of their predicament? "They're certainly sen-tient," Mills insists, and he launches into what I am sure is his stump speech. "I think most people at heart are compassionate toward others, be they people or nonhuman animals. About thirty thousand species of plants and animals are going extinct every year in this country right now, just because of us. So we're going to eat a species off the planet that has been here for two hundred million years?" His voice rises. "For aphrodisiacs, for a better sex life, for food?" Totally unnecessary, is his answer to his own questions. "I come from a culture of slavery and child abuse and incest and wife beating," he says about the South he left. "Some traditions deserve to die." He rejects that his concern for turtles is at the expense of ancient Chinese traditions. "That justifies nothing. A lot of things have been around for thousands of years that need to go by the wayside." Yet for all his work on behalf of turtles, Eric Mills claims no special affinity for the animals. "No more than any other animal. I think every species is special." But he does know what influences public opinion. "If these turtles and frogs were half as cute as a panda, we wouldn't be having this conversation today."

The next day I return to Sun Hop Fat 1 for another look, and the pile of turtles seem even worse for the wear. Only a couple of the turtles are crawling around; one of those in particular looks intent on trying the impossible: escape. A chatty customer asks me if I've eaten turtle soup. "It's good," he says encouragingly. "Tastes like seafood." Blue Smock is on duty, watching us, as my turtle soup aficionado informs me that the turtles he buys are killed in the store.

In the fall of 2014, a few years after the live-turtle trade resumed in California, a guide on the open upper deck of a tour bus driving down Stockton Street through San Francisco's Chinatown added to the rancor. It was her last day on the job when passengers were treated to a unique spiel as she railed against the neighborhood where she once lived with a torrent of ethnic abuse over the bus's loudspeaker. "Fuck your little seafood fucking markets with your turtles and your frogs inside, okay?" she yelled at Chinatown. "When you come to America you gotta assimilate a little bit, and here in America we don't eat turtles and frogs. But they gotta bring that here to America," she ranted. "There's a limit, okay? You gotta assimilate a little, Chinatown." When the manic guide took a breath, some of the passengers applauded, and as the bus entered the Stockton Street Tunnel she encouraged them to chant, "Fuck Chinatown." As soon as he heard about her good-bye verbal assault, City Sightseeing CEO Christian Watts apologized, promising new training standards for guides so that "each tour is up to our rigorous standard." *NBC News* posted a video of her profane sermonette, complete with the caveat, "Warning: Some viewers may find the language used in the video disturbing."[9]

— *Soup's On* —

Here in America, of course, we eat turtle; our taste for it goes back to precolonial times and transcends Chinese and Chef Folse's kitchens. Turtle soup is still on the menu at Commander's Palace in New Orleans, the restaurant that tags its offerings "haute Creole cuisine."[10] Delmonico's—which calls itself America's first restaurant—no longer carries turtle on its menu, but when it did, Oscar Wilde reputedly was thrilled with the taste and wrote, "The two most remarkable bits of scenery in the States are undoubtedly Delmonico's and the Yosemite Valley."[11] Delmonico's Chef Alessandro Filippini comes right to the point in his cookbook regarding technique, and he's not worried about the turtle's feelings. "Take live terrapin," he instructs, "and blanch them in boiling water for two minutes. Remove the skin from the feet, and put them back to cook with some salt in the saucepan until they feel soft to the touch. Remove the carcass," he continues his instructions, "cut it in medium-sized pieces, removing the entrails, being careful not to break the gall-bag." Apparently, turtle gallbladder tastes bitter. Next the cut meat goes into another saucepan, "adding two teaspoons of pepper, a little nutmeg, a tablespoon of salt and a glassful of Madeira wine. Cook for five minutes, and put it away in the ice-box for further use."[12] For example, in Chef Filippini's famous terrapin soup.

Just as brutal was a recipe published by the Culinary Arts

Institute of Chicago in 1947. "Drop live terrapin into boiling water," and after five minutes, "remove from water; rub skin off feet, tail and head with a towel, drawing the head out with a skewer." This would be an ideal time to drink a glass or two of Chef Filippini's Madeira. "Clip off claws. Scrub shell with boiling water; break apart with a cleaver or axe. Remove meat and liver. Discard heart, sandbag, entrails and gall bladder. Cut the liver in thin slices." This candid recipe also warns of the bitter gallbladder. But salvage any eggs found. "Take out eggs, remove film and set aside in cold water. Mash yolks; add flour, nutmeg, lemon juice and rind." Next comes a cup of soup stock. Then add some onion and celery to the meat and eggs and cook the concoction "until meat falls from bones." Take out the bones and "serve with toast."[13]

Turtle conservationist Peter Paul van Dijk, the scientist convinced that turtles in his care recognized him when he joined them in their greenhouse lair, acknowledged eating turtle when we chatted at the Turtle Conservancy's compound in California. It was lunchtime on a turtle farm in China, and he decided it would be too rude to refuse what was on offer. Van Dijk picked up his spoon and slurped. The man loves turtles and respects individual lives, so how did he cope with a spoonful of turtle? Emotional detachment is his tactic. "You just have to kind of separate yourself from it," he advises. "Just like when I walk through a market in China and see somebody buying a turtle and they request, 'Can you just parcel it up into pieces for me?' You have a turtle that goes from live animal to a bag of bits in fifteen seconds and, yes, it's a life that's being snuffed out." He sighs.

"How did it taste," I ask, of course.

"Like the sauce."

"It could have been tofu," I suggest.

"It could have been tofu," he agrees. "It would have been just as tasty." But he understands the lure. "It's not about the taste. It's about the perceived medicinal value, the perceived exclusiv-

ity and conspicuous consumerism. There is absolutely no reason to eat a five-hundred-dollar bowl of tiger penis soup except
to show that you can treat your friends, your guests, to the most
expensive meal available."

— *Poaching for Salsa* —

Downtown Sámara, Costa Rica, on the Nicoya Peninsula along
the Pacific coast, is a few dusty blocks of cheap lodging, noisy
pickup bars, and shops selling the requisite *pura vida* T-shirts
and straw hats. The town remains just a burdensome enough
drive from San José to keep it from becoming completely over-
run with *turistas* despite its idyllic palm-lined white sand beach.
But for tourists who do make it to Sámara looking for more than
drink and surf, local guide Roberto Brenes offers a short trip
south along the coastline to Camaronal and an opportunity to
watch hatchling sea turtles scurry into the ocean.

I hire Brenes with an ulterior motive, figuring if I go on one
of his standard turtle tours and we hit it off, he might be an
entrée to the local turtle egg poaching scene. Along with a couple
of tourists from Europe I pile into our guide's four-door pickup
heading for Camaronal. The bucolic beach is typical of the Central American coastal paradises where olive ridley sea turtle
eggs are deposited in nests made by the females during their
dramatic mass *arribadas*. The tour is delightful, even inspira-
tional. Our guide parks the truck and we hike about ten minutes through the jungle. We see a toucan with its Jimmy Durante
beak and bright yellow breast. A family of howler monkeys looms
high in a guanacaste tree, howling as we tramp past termite
mounds, howling and looking down on us with what sounds to
me like threatening disdain for trespassing on their turf. Bright
yellow butterflies flit along the trail. The beach is another perfect one, the sky and water a deep tropical blue.

Student hatchery workers in shorts and flip-flops meet us. Their job is to gather eggs as soon as the mother covers her nest and heads back out to sea. They stash them in a beachside hatchery where they incubate safe from poachers and predators—raccoonlike coatimundis scavenge eggs and crabs grab hatchlings on the beach. This day's olive ridley hatchlings are ferried from the hatchery to the beach in a weathered Styrofoam beer cooler. One of the students tips it sideways at the tide line onto the black sand. The turtles—about the size of those that used to be given out routinely as county fair prizes—scurry out.

The Camaronal turtles skitter down the beach toward the breaking surf, most traveling in direct paths to the sparkling Pacific. The lapping waves dump them askew, and with their flippers they relentlessly soldier on into the ocean seeking deeper water. A few stragglers look tired and confused. Students grab them and point them back toward the surf. Where's their mama? They'll never know. One staggers, dragging a damaged front flipper.

"Do you think he's going to make it?" I ask one of the students.

"No, but he's fighting." Whatever the life force is that keeps us all going is on vibrant display as the hatchlings—even the lame one—keep going toward their destined ocean home.

"Come on!" encourages a student worker. "You can make it!"

Our guide lives near the beach in the Camaronal jungle and invites us for refreshments at his home, a house he built himself with native woods. The energetic twenty-two-year-old is still at work on a precarious-looking gangway heading a hundred feet up into the trees, "so tourists can see the ocean." Brenes studied business and sustainable tourism at the Universidad Nacional in Nicoya. His turtle expertise is self-taught from on-the-job training. What looks like it could be an incipient Vandyke beard is growing on his Vandyke-brown chin. His smile is welcoming. We trek back to Sámara, say good-bye to the Europeans. The counterpoint to the lovely scene we witnessed this day is eggs stuffed into poachers' sacks and sold to beachside bartenders who mix their under-the-counter potions for sex tourists. I tell Brenes that I want to commission him for another tour—a tour to meet these poachers and bartenders.

It's dark and the Sámara bars are cranking up the dance noise as we head out of the village back toward Camaronal and the jungle house of an old-timer Brenes knows, a subsistence turtle egg poacher. The local poachers, he tells me as we bump along the rugged road, gather the eggs for their own food and to sell to neighborhood bars for a few *colones*. Reggae is blasting out of the truck's speakers. Between stories, Brenes sings along. At the bars the eggs are added to salsa to make the alleged aphrodisiac. "I tried it," says my guide. "I don't like it." Volunteers patrol the beaches in an effort to keep the poachers from the eggs, but they patrol without police powers. Wardens are assigned night duty to Camaronal and the other beaches near Sámara frequented by egg-laying turtles. "The three rangers," shrugs Brenes, "usually are sleeping." He takes a utilitarian approach to the poachers. "It's really hard to tell them not to poach when there's no alternative way for them to make money." He admits a grudging respect for their technique. "Getting leatherback eggs requires grabbing them just as she's laying them. Otherwise you'll never find them." The mama covers her nest so its whereabouts are camouflaged with sand. "For successful poachers, it's an art." He reports a live-and-let-live status quo among the turtle huggers like him and the egg thieves. "I know them, but I don't try to confront them." That attitude can be a survival tactic in Costa Rica.

Across the country on its Caribbean coast Jairo Mora Sandoval was murdered on Moín Beach during a nighttime patrol checking on nesting leatherbacks, trying to protect them against poachers. The coastline there is notorious for its drug traffickers, gangs that augment their incomes and diets poaching turtle eggs. Threats against Mora, a well-known and vocal sea turtle conservation activist, were frequent. "He was held up at gunpoint and told to stop the walks," fellow advocate Vanessa Lizano said after Mora was abducted, tortured, and killed in 2013.[14] Three years later, their acquittals in a previous trial overturned, four gangsters were convicted of murdering Mora and sentenced to maximum prison terms of fifty years. Chief judge Carlos Álvarez cited Mora's turtle protection work as the killers' motive and the murder as a national blight. "This crime is more than just a horrible murder," Álvarez said as the trial ended. "It has also damaged Costa Rica's reputation as a green country. It has scared away environmentalists."[15]

Not all turtle egg collecting is illegal. Just a few hours north of Camaronal is the Ostional Wildlife Refuge, famous for its olive ridley *arribadas* and their relatively easy access for curious tourists. It's legal for locals to collect a limited number of eggs from the first wave of nesting turtles, eggs often otherwise destroyed—trampled by later arriving females.[16] But with only a couple of park rangers assigned to guard the beach, enforcement of the restrictions is wishful thinking.[17] Worse, the legal allowances at Ostional fuel the illegal trade by whetting appetites, and those poachers who worry about the authorities can equip themselves with counterfeit receipts that claim their eggs were collected legally.[18]

Brenes pulls his truck up in front of a small plain house. We needed its four-wheel drive to ford rivers and survive the dirt tracks posing as roads. The dark night sky is peppered with shimmering stars. Guard dogs bark. Brenes reassures them, and we pass through the gate into the yard. An old man with terrible-looking teeth hobbles out to meet us, complaining about his hurting leg. His baseball cap reads FREEDOM in English. He knew we were coming; Brenes had called ahead. Inside the house I see a woman I presume is the man's wife, busying herself with housework while eying us with understandable curiosity. We three men sit in the yard and talk about making a living, treatment for the leg, and turtle eggs.

"There was no other option," the old man says about the days when he collected and ate turtle eggs. "We would mix them up with some lard, sauté them, and eat them in sandwiches as if they were chicken." But such meals no longer are an option for him and his family, he claims. "It's wrong," is his simple response to poaching, now that he understands the turtles' fragile existence. But it's not just his conscience; it's also his body. I ask him if he would dig for eggs if a poacher paid him for his expertise finding nests, and he points to his game leg, the legacy of a soccer-playing accident. "Look at how I am now. No way. I can't. Before, I used to sell them. That was the reality." He is not judgmental regarding those neighbors who feel they must continue to collect eggs to survive. "They feel obligated," he says, "and that's okay when it's for eating and there is no other option."

Brenes goes out the gate to the truck and fetches a bottle of rum. "For the leg," he tells the grateful hurting old man. We leave the poacher and bounce along the dirt road for a few kilometers to La Costa, a tin-roofed open-air roadside saloon decorated with Costa Rican flags on a string like Tibetan prayer flags, a huge TV screen, and light bulbs in wine bottle shades glowing amber. "This is where the salsa is sold," Brenes tells me as we're greeted with more reggae.

The proprietor claims the potent stuff he sells is legal because he gets the eggs from Ostional, sustainably harvested and with permits. Roberto Brenes and I belly up to the bar and order the "special sauce" from the *mamacita* who makes it, and we ask for the turtle-free salsa. She takes a bottle from a cold case and pours the red viscous stuff into shot glasses for us. "What's in it?" I ask, holding my glass up for a toast.

She recites the ingredients: tomato juice, onion, cilantro, rosemary, chili—both *dulce* and *picante*—salt, and *mucho cariño*—which could be much care or much affection or much love, depending, of course, on the moment.

"And turtle?"

"No *tortuga*," she guarantees.

The next day I'm in the Hertz shuttle to the San José airport stuck in an insane traffic jam that the driver is slaloming through like an Olympic skier. I congratulate him and tell him he is an *artista del traffic*. We pass a truck filled with pigs, and he says he loves pork at Christmastime. Not me, I tell him. Are you a vegetarian, he queries. I eat fish. How about *tortuga*, he asks. We talk about sea turtles, about their endangered status. Eating *tortuga* makes you strong, he announces, grabbing one of his biceps. He talks reverently about seeing masses of turtles on a beach during an *arribada* and of watching hatchlings going out to sea. "Years ago I ate turtle, but no more," he says. "You can go to bars and they'll serve it. But you can't know if it's legal or not since there are legal harvests." We're still stuck in the airport-bound traffic as Mr. Hertz describes how tomato-based sauce masks the strong taste of the turtle egg in the potions served to the special speakeasy clientele—which makes me wonder: Did I drink a shot of turtle egg last night?

• **Cousin Fred** •

Sheila and I find ourselves talking to and about Fred as if he's one of the family. "How's Fred?" I ask when I come home from the university. "Fred's not interested in the lettuce," she replies. When I walk past his house, I peer down and hail him with a hearty, "Hello, Fred." Reaction, if any, is slo-mo. I look him in the eye and try to determine if he's annoyed, checking me out, or uninterested. I begin to study and appreciate his complex parts: old-man head and retractable extremities, wrinkled flesh and domed shell, claws on his feet and little pointed tail.

Meanwhile, Sheila and I talk about Fred.

"What's Fred doing now?" she asks.

I take a look and report, "Not much."

"Is he back on his heating pad?"

"Yup," I say, and suggest he's doing okay by referring to him by his full name, "Good ole Fred."

"Good ole Fred," agrees Sheila.

— *The Government's Turtle Stash* —

I'm in the National Eagle and Wildlife Property Repository out on prairie land close enough to Denver that it's surrounded by the city's eastside urban sprawl. This place was the Rocky Mountain Arsenal, established during World War II to develop chemical weapons such as the napalm dropped on Tokyo to start firestorms. Rocket fuel was also produced here, as were pesticides. By the early 1980s the arsenal, a poisoned dumping ground for the by-products of the poisons it manufactured, was deemed obsolete. After a roost of bald eagles was discovered in a cottonwood grove on the former base, a massive cleanup job transformed the landscape into a wildlife refuge. Here, in its vast windowless warehouse, the Fish and Wildlife Service stashes confiscated smuggled animal parts. Shelves in the repository are piled high with all kinds of turtle things. Walking the aisles is like being in a Costco of contraband.

My guide is Coleen Schaefer, the repository's resident supervisor. Her vibrant red hair hangs shoulder length to her brown FWS uniform cardigan, a look that softens her official stance, as do her warm brown eyes. We're in the turtle section, an aisle or two beyond the shelf where a quartet of severed leopard heads rests and a wooden hanger holds a full-length leopard-skin coat topped with a matching leopard-skin pillbox hat (and how *would* your head feel under something like that? asks Bob Dylan in "Leopard-Skin Pill-Box Hat" circa 1966). Schaefer shows off the stock, initially sounding as much like a shopgirl as the cop she is. "We have traditional Chinese medicine," she shows off a clear plastic bag marked "turtle powder." Inside the evidence baggie is a package labeled "healthy food" under an illustration of a turtle along with instructions for making the powder into a "popular" soup. "Here's a carapace made into a mask." Schaefer displays the mask, the mouth, nose, eyes, and mustache fashioned in hammered silver. She picks up another dead turtle and studies it, trying to figure out its purpose. "It's a glass flask!" she says with surprise after finding the neck of the bottle and the gold-colored screw cap made to look like the dead turtle's head. She almost sounds delighted—it's ingenious and intriguing—before remembering it was a turtle before it was killed and turned into a frivolous accessory. "Disturbing," she says, returning it to its display place. "Here's a carapace that was made into a purse." Leather is laced into the shell to form a pouch, an ideal stash for turtle powder. "This is herbal tortoise jelly." Schaefer is holding a jar decorated with Chinese characters and identified as a product of Wuchou Manufacturing Chemist in Kwangsi, China.

What's in it? She reads the label. "'Herbal Tortoise Jelly is prepared with more than ten chosen valuable Chinese medicine with scientific method. Having desintoxicating, GIT regulating and blood nourishing effects, it is a special tonic to health maintenance, suitable for adults and children.'" Schaefer shakes it and pronounces, "I'm sure it does all that and more," her sarcasm clashing with what's been a matter-of-fact rendering of the collection she supervises. "Disturbing," she says again.

Whether a turtle product is legal or illegal depends on the species and, for those species legal to trade, whether the buyers and sellers hold required permits. "Without a permit we have no way of knowing if it came from a sustainable source," says Schaefer as we continue our tour. The repository takes custody of confiscated wildlife once a case is adjudicated and the turtle remains are no longer needed as evidence. About a million and a half animal parts are stored at the facility, kept for potential service as show-and-tell props. Nothing beats stark examples like the leopard heads when educating the public. Whatever Schaefer and her crew do not need for lessons is destroyed; what's in storage is only a fraction of what inspectors catch at U.S. borders and, of course, does not reflect what is missed and crossed into the country undetected.

"Who wants to buy this stuff?" I ask Schaefer. What drives

the illicit trade? Who wants a model of a ship decorated with tortoise shell sails sitting on their mantel?

"I think about that often," says Schaefer, looking at the stacks of turtle paraphernalia. Before taking on management of the repository, she was a wildlife inspector working with customs officers catching smugglers, and she knows the type. "Huge egos. If they're interested in drugs and money and arms, then they're going to be interested in something unique that somebody can't have." Endangered species fit that demand. "If you're looking for drugs, you're going to find endangered species because they're rare and they're valuable and they're unique." Wildlife crime, says Schaefer, is no longer just opportunistic crime. Organized gangs now recognize that there's money to be made smuggling animals and animal products; it's a low-risk, high-reward business. For those who are caught, prison time is usually minimal, so even when smugglers are caught, "of course it's worth it to them. They're making millions of dollars and they can expect a sentence of just a couple years." The threat posed by wildlife and animal product trafficking is more than concern about endangered species, she says. It's a security issue. Insurgents in Africa are selling poached elephant tusks and rhino horns to bankroll their violent acts.

The tour continues. Stacks of stuffed sea turtles. Wooden cabinets decorated with inlaid tortoise shell. A stringed instrument—it looks like a banjo—with a carapace for the body. Piles of mottled brown and tan combs with broad filigreed handles, the type Victorian women sported atop their piles of hair.

Schaefer's training is as a wildlife biologist, a duck-hunting biologist who raises chickens and pigs on her farm—and eats them. "I find it respectful to use every aspect of an animal if I'm going to harvest it. Otherwise I find it wasteful." I point at the combs and suggest one could make the same argument about the sea turtle. Perhaps it was eaten and now its shell is put to creative, practical, and permanent use. "You definitely could," she agrees, her voice quickening with enthusiasm for the question before she adds, "but my values for wildlife are very species-specific. I can have this utilitarian value for my pigs and my chickens and ducks because I'm an ethical hunter. It does not translate," she says with finality, "to all other species." Of course, it's perfectly legal to wring a chicken's neck, slit a pig's throat, and blow a duck out of the sky. Making combs out of sea turtle shells is not. Zebras and elephants and rhinos and turtles she considers more precious than farm animals and abundant waterfowl because of their scarcity, because they are under attack from poachers, and because their habitat is being destroyed.

Another aisle. Boxes and boxes of cowboy boots made out of sea turtle skin in a rainbow of colors. Turtle skin pumps and turtle skin high heels. Purses, belts, and wristwatch bands. A box labeled turtle oil soap made in Mexico promises "softness to your skins [sic]."

Supervisor Schaefer tries to understand those who use illegal animal parts but are not obsessive collectors. Furs and skins keep people warm and come from seemingly renewable sources, while manufacturing synthetics generates a nasty carbon footprint. "There's a lot of validity to that argument," she says. "I lived in Alaska and worked with the natives up there. They're not going to wear synthetic around their faces. They're going to wear wolf, they're going to wear seal. They're going to wear what's worked for them for generations."

More shelves. Pills, salves, and what's just marked "turtle medicinals." Ornamental butterfly figurines. Another stringed instrument, this one with a butterfly image in tortoise shell inlaid between the bridge and the sound hole. Turtle heads mounted for wall decoration. Stacks of sea turtle hides.

Does any of it appeal to Schaefer just as curiosities? "No," she says without hesitation, "especially with sea turtles." She began her career patrolling Mexican beaches, watching olive ridley turtles laying eggs, "knowing that out of a hundred there's only going to be one that's going to make it."

More: rings, earrings, pendants, and pins. Bracelets. Hairbrushes. Mirrors. What if somebody's grandmother offers her granddaughter a gorgeous tortoiseshell hairbrush and comb and mirror set that's been in the family since pioneer days or migration from Europe? As long as it predates the Endangered Species Act and is a gift—not sold—antiquities made of turtle are legal to keep.

We walk back to the classroom Schaefer and her staff use for lectures to visitors. There I meet Rudy, a mannequin decked out in a pair of Clark Kent–style glasses. The frames are tortoiseshell, of course. Which reminds me that long ago a family friend gave me his pince-nez, and as I look at Rudy I remember they're tortoiseshell—and so old perhaps they're real tortoiseshell. They're back in my home office, at the bottom of a drawer. When I return there I must check them out (and be sure not to wear them around Fred).

— *Tortoiseshell Guitar Picks* —

But the pince-nez is not the only tortoiseshell oddity that's infiltrated my family. A colleague from Central America and his wife, as a gesture of affection, presented a gorgeous amber-colored bracelet to Sheila: a bracelet, we realized quickly, made of tortoiseshell. What to do with such a gift? One of my sons, a guitarist, is awed by what he considers unique sounds that he attributes to playing with a tortoiseshell pick.

"I grew up hearing stories about tortoiseshell picks and their magic qualities," says Tal, who's played all over the world with a variety of bands. "It's legendary," he says about the picks, and once he played with one, he understood why. "Without question. From an acoustic guitarist's perspective, the variety of tonal possibilities you can get with your picking hand using a tortoiseshell pick far outstrips almost anything that's out there on the market. When you're dealing with the expressive power of getting your point across musically—especially with a recording where the microphone is basically a microscope for sound—the tortoiseshell pick allows you a much larger amount of control over your sound."

Why would it make a difference, I wonder? He showed me a tortoiseshell example. Like all the tortoiseshell commodities I've observed, it was gorgeous—translucent, mixed with dark patches that reminded me of a Dalmatian's coat, or maybe serious summer storm clouds over Oklahoma. But a pick is a pick from my perspective since I'm no guitarist.

"To a guitarist, the pick is the bow," he says, comparing his instrument to a violin. "The pick is the actuator. It's the point at which your body is contacting the string and creating rhythmic information. When you attack and release the pick on the string, there's a particular moment when it's actuating the vibration on the string. Some pick materials accentuate the treble frequencies. Some, like a felt, accentuate more of the bass. The tortoiseshell pick is a balanced attack, so you are getting almost everything the guitar has from treble to midrange to bass." It is in expressing the midrange that the tortoiseshell excels, explains Tal, frequencies difficult to obtain with a plastic pick. "It's a very balanced and powerful sound." His enthusiasm for the prowess of tortoiseshell picks continues until I interrupt him, suggesting that the nuances he seeks escape my ear.

Not that Tal favors killing turtles and tortoises to achieve what he considers an ideal sound from his guitar. "If someone were to tell me that a tortoise was hunted, a tortoise was slaughtered, and the result of that violence was what I held in my hand, I would find that aberrant and I would not want to use it." He doesn't seek the picks, nor does he buy them. But he's ended up with a handful that came to him the same way the bracelet arrived, as gifts. Other guitar picks are sold to musicians fabricated from repurposed tortoiseshell: antique combs and hairpins and tabletops and other artifacts of the heyday of tortoiseshell as a material to make various products. Tal's pleased that a new

Primetone synthetic tortoiseshell pick all but replicates the special qualities he finds with the natural ones. Meanwhile, he's comfortable with the tortoiseshell picks he uses because "I was told they come from a farm. If I'm opposed to that, then I'm also opposed to chicken or hamburgers."

I hate to wreck my son's daydream about politically correct tortoiseshell artifacts carved out of sustainably raised hawksbill turtle shells. But there is only one sea turtle farm in the world and it's famous for green sea turtles (*Chelonia mydas*) and controversy, not hawksbills and guitar picks. Founded as a for-profit operation, the Cayman Turtle Centre is now owned by the government and operated by the Ministry of Tourism. Its promotional material shows happy, sun-drenched tourists swimming with and manhandling green sea turtles. But the Sea Turtle Conservancy, among other turtle protection organizations, opposes the center's exploitation of farmed turtles for their meat and shells and opposes the government's desire to export farmed sea turtle products. "Reopening of trade," says the conservancy, "will create the kind of demand that fuels illegal poaching and black markets for turtles caught from the wild." Conservationists also deplore the center's breed-and-release program, worried that the farmed turtles will breed with the wild, a recipe for spreading disease and confusing migration patterns.[19]

There is a guitar connection to the Cayman Turtle Centre: Paul McCartney lends his name to the World Society for the Protection of Animals campaign to shutter the place.[20] The literature from WSPA's campaign paints a nasty picture. "Sea turtles are solitary, migratory creatures," it notes. "Imagine the cruelty of trapping one in a small filthy tank." Multiply that image by thousands to understand the savagery of the center. "They get sick and stressed. They turn on each other—biting and maiming one another." WSPA lobbies the Cayman Islands government to convert the center into a rehabilitation operation.[21] Meanwhile, the center lures tourists with lines like "Reach in to touch or pick one up for a photo of a lifetime!"[22] And after such a frolic, the hungry tourist can further experience a center turtle at one of the Cayman eateries featuring farmed and legal—though morally questionable—sea turtle stew or sea turtle steak.[23]

Despite the controversies surrounding the Cayman turtle farm, turtle ranching worldwide is a thriving business.

• Contented Fred •

I'm a little worried about Fred. Maybe a little more than a little worried. He's not eating much at all. He ignores a second scoop of fancy Whole Foods dog food. He expresses no interest in lettuce. He pokes at and maybe takes a few bites of some carrots Sheila shredded for him. When I take him out of his box-house he mostly just sits; the running all around the living room floor that he seemed to enjoy his first day in Oregon is a distant reptilian memory for us both. He seems to be drinking—or is the heat from his basking lamp causing the water in his dish to evaporate? There is no evidence ole Fred has defecated again. When he rejects a slice of banana (just mindlessly steps on it as he walks toward his water), I decide to offer him another worm.

At first Fred ignores the worm lying in his dish on the bed of shredded carrots and lettuce. But when the worm stops playing dead and wriggles, Fred makes his move—pecking at it with his beak, sometimes grabbing it and shaking it around. He acts like a cat with a mouse; it seems he is just playing with Mr. Worm. But it's more than a toy. After a few minutes Fred takes a worm end in his mouth and starts to swallow. The worm begins to disappear into Fred's mouth like spaghetti, albeit spaghetti that's fighting for its life as it flails about, making Fred work for his supper. It takes a few minutes for Fred to dispatch most of Mr. Worm, and when there's just a stubble left, the worm no longer moves. Before Fred takes his last bite, the stiff worm end sticks out of his mouth reminiscent of an unlit cigar. Fred looks like a well-fed Bürgermeister who's retired to the smoker after dinner in the dining car. Dare I say he appears content?

FIVE

THE PRODIGIOUS FARMS

— Hoovering Turtles —

"We have seen the Chinese trade vacuum out one region after another," turtle conservationist Peter Paul van Dijk says about China's rapacious appetite for all things turtle.

This vacuuming of turtles exacerbates the natural vulnerability of turtles. Even without man-made threats and collateral damage, the meager survival rates of hatchlings combined with the many years it takes most species to reach an age when they can reproduce, make turtle species extraordinarily susceptible to extinction. The deleterious effects of a turtle species becoming extinct both as nontangible components of cultural heritage and as active participants in a thriving ecosystem are real, consequential, and permanent.

— Turtles Breed Crime —

Lust for turtles in Asia is not just fueling illicit turtle trade; there's plenty of legal activity. Chen Xingqian is a Chinese turtle farmer who saw an advertisement in 1981 with the promise, "Raising turtles can make you a fortune." According to his rags-to-turtles story, he started with six; today he breeds over one hundred thousand a year. He is now worth millions, thanks to his success-ful turtle farm and the sales of its stock.[1] Most flamboyant of the Chinese turtle tycoons is Li Yi, whose stash of golden coin turtles (*Cuora trifasciata*) numbers upward of twelve thousand, according to Western turtle experts who have visited his farm. His wealth must be staggering, given that each of those little guys is worth about $10,000. These exceedingly rare box turtles are highly prized, not just for their beauty but also for their pur-ported medicinal qualities. But this risky big-bucks animal business attracts criminal human predators.

Turtle magnate Li Yi—a policeman before going into the tur-tle business—protects his operation with armed guards, mean dogs, and CCTV cameras. His golden coins breed and await their fates in an eight-story building designed to resemble a turtle, a luxurious Club Med for both turtles and ticket-buying turtle tourists. He sells live turtles, but he also grinds his stock to make and sell his high-priced brand of turtle jelly. Expect to spend as much as $500 a dose for the fancy stuff.[2]

Golden coins are all but extinct in the wild; their last known natural habitat is in Hong Kong's New Territories. They are pretty turtles, with yellow coloring on the edge of their carapace, yellow on the back of their head, and orange lower legs. Accord-ing to Chinese tradition, that golden hue is considered good luck—but not so much for the turtle. When it's kept for luck, at least it's kept alive. Unfortunately, it's also sought after in tradi-tional Chinese medicine as a cure for cancer and most every-thing else. Its shell is a component in *guiling gao*, turtle jelly, a compound sold the way snake oil was in America's Wild West: good for whatever ails you.

One outfit making grandiose claims about its turtle jelly is Hoi Tin Tong, founded in Hong Kong, company literature proudly claims, by Mr. Ng Yiu Ming, a purveyor of turtles and other wildlife for medicinal purposes. "Mr. Ng's mother weighed less than 100 pounds during sickness," a brochure explains, "but after taking Hoi Tin Tong Herbal Jelly consecutively for quite a period of time, her health has improved dramatically and she can eat normally with her weight getting back to about 140 to 150 pounds." Mr. Ng responded when "everyone" wanted to try his jelly by opening a shop on Bowring Street, eventually adding outlets in other Chinese cities.[3] The business faced a crisis in 2013 when a professor from the City University of Hong Kong tested Hoi Tin Tong jelly and found no evidence of turtle in three out of four samples. Mr. Ng rejected the results, saying, "We use fresh turtle for production and boil the whole turtle, including its shell and belly, with herbs."[4] He dismissed the test results, insisting that the amount of turtle in his jelly is not what makes it work: the secret is his formula. Mr. Ng is a TV personality in China; he is featured in his turtle jelly commercials feeding a happy baby and showing off his healthy turtle supply— healthy, at least, before they become jelly.

— *Indonesian Turtle Bazaars* —

Jakarta. Notorious for wide-open animal trafficking in some neighborhoods and perpetual traffic jams almost everywhere. The Sultan Hotel lobby door whisks open and my air-conditioned self is whacked by the outside heat and humidity. In a minute I'm in a taxi so cold I ask the driver (who is also my guide) first to take the chill off the AC and then to take me to Jatinegara, a neighborhood infamous for its live-animal market: birds, mammals, and reptiles.

"Maybe it's not good to go there," he says with gracious diplomacy. "Could be dangerous. Not very safe."

I acknowledge his concern and once more ask him to head for the market.

"You do not want to go there," he says again. "It's a bad place with bad people. Maybe they take things." He points at my Nikon.

Nonetheless I repeat my request, but temper it with what feels like a fair caveat. After all, he's not the only cabbie in the city. "I do want to go to Jatinegara, but please take me only if you feel comfortable there."

He shrugs an okay. "I live there," he says, as we inch into the ludicrous and frustrating Jakarta stop-and-go.

But no turtles or tortoises are evident when we survey the chaotic marketplace that stretches along a main road and off into the distance along a ramshackle narrow alley of shops. Heartbreaking cages stacked high pack the sidewalks, tiny cages jammed with birds and cats and forlorn-looking monkeys all awaiting their fate. The monkeys grasp the bars staring at the outside from their jail cells. I snap a picture as their captor bellows an angry "Argh!" at me. Time to leave.

Back into the river of traffic the taxicab flows, swimming slow motion among cars and trucks and motorbikes. Next stop in another shabby neighborhood is a shop called Severa Reptile—the name sounds promising for finding turtles. I walk past a cage crowded with dirty mice, their fur filthy and matted. The white mice await their fate next to big aquariums filled with thick, undulating snakes. It's a messy store, strewn with critter food and supplies. The clerks ignore me as I look around, voicing no objection when I take out my camera.

Opposite the snakes, big plastic tubs litter the floor. I peer into the tubs and there they are: turtles on offer, their inventory numbers painted in white on their carapaces. I shoot pictures while a clerk I summon provides prices. There are plenty of gorgeous Indian star tortoises (*Geochelone elegans*), 3 million rupiah each—about $250 in U.S. dollars. I spot three pancake tortoises (*Malacochersus tornieri*), so named because they are as flat as flapjacks, an adaptation that suits their habitat: narrow crevices in rocks. They fetch about the same price as the Indian stars. And in a deep blue tub, the vibrant colored plastic a heavenly backdrop for the glorious golden pink starbursts exploding on its pitch-black carapace, is a lonely radiated tortoise (*Astrochelys radiata*), available to take home for a cool 20 million rupiah—over \$1,500. I leave passing cages of chameleons that must question what color is best for surviving their immediate circumstances.

That evening I email the photos to the Turtle Conservancy to confirm their identities. Eric Goode writes back with depressing commentary about the overt trafficking I'm witnessing in Jakarta. "Indian Stars are wild-caught and smuggled out of India in huge numbers," he reports, and then queries, "Sure they are Indians Stars and not Burmese Stars?" The question is germane because the Burmese stars (*Geochelone platynota*) are even more rare than the Indians. The pancake tortoises he notes as "critically endangered from Kenya or Tanzania." And that 20 million rupiah specimen? "The tortoise in the blue tub," he confirms, "is a critically endangered and fully protected Radiated tortoise from Madagascar."

Again into the river of traffic we flow, the endless motorbikes swarming like mosquitoes around the miserable jam, north toward the Java Sea and the Mangga Dua Square shopping center. I get out of the cab at the wrong corner and risk my neck dodging cars and trucks and motorbikes—some of the noisy, smoke-spewing bikes hauling three and more riders. Vendors peddle unidentifiable (by me) street food from Technicolored wooden carts. Drivers lounge on their *bajaj*—the three-wheeled motorized version of pedicabs.

The mall is a massive building sprawling across several city blocks, a complex that's probably seen better days—hard to imagine it looked so tattered when it was new. Inside, past the Baskin-Robbins ice-cream stand, a crooner is bellowing karaoke in a rhythmic Bahasa Indonesian, or at least that's what I presume. My own skill in the language is limited to *penyu*—turtle—a key word for this trip, along with the standard opening-gambit vocabulary: hello, thank you, and my favorite: *hati hati*. I learned it from road signs. It means "watch out," "be careful." Words to live by on the noisy, crowded, and crumbling streets of Jakarta. Across from the singer is a kiosk, and its shopkeeper beckons

me with a Taser. When I decline he waves me toward binoculars. I pass on the spyglasses, the ice cream, and the down-market
clothing (offered at a 70 percent discount) and head for what's
billed as an exotic pet show.

Kiosks, a dozen or more, are jammed with an array of reptiles, including turtles. Tortoises wait in plastic tubs, and turtles
try to swim in Tupperware so small there's nowhere for them to
go. I'm offered what's identified as a common snapping turtle
for about $550, "from the Amazon," says the seller. More likely
it's bred in China from American stock. At another booth the
snappers are from the United States, according to their purveyor, shipped to China, and then on to Indonesia. Another
says his snappers are bred in Hong Kong. Still another holds up
an alligator snapping turtle about the size of his palm. Just a
baby, yet it already displays its aggressive temperament with a
fierce-looking snarl. Several North American species in demand
by Asian buyers are now bred in China. In the shopping mall is
another radiated tortoise. I ask to photograph it. "No pictures,"
says the owner as he hides it from view. "It's not for sale. It's
mine."

Turtle conservationists tell me Indonesia takes a laissez-faire
approach to the international treaties the government signs,
agreements that are designed to protect endangered and threatened animals from the marketplace. If authorities enforce anything, it's the laws against trade in species native to Indonesia.
Chelonians from Madagascar or Burma or elsewhere warrant
only a yawn.

Sad but true, agrees an investigator working in Indonesia for
a premier international organization that fights illicit and im-
moral wildlife trafficking. Given his clandestine lifestyle ("I'm
not supposed to be here, and there's the personal safety aspect
as well. I've never had any problems, touch wood."), I'll refer to
him as James Bond and not describe his appearance or his
Indonesian redoubt.

"The main problem," Bond explains, "is that the Indonesian

law doesn't protect non-native species." It's a loophole that he and other wildlife preservation activists are working to close. Meanwhile, he laments that what I saw is routine: it's open season in Jakarta for trade in critically endangered species as long as they hail from elsewhere, and he's seeing more and more North American species in the Southeast Asian markets. Although laws are on the books protecting species native to Indonesia, "law enforcement is still greatly lacking and one of the main setbacks is the level of corruption here."

James Bond does find allies in Jakarta, Indonesians struggling against bad odds to save turtles and other animals traded in the shops and street markets. "They're proud of their animals," he says about the merchants. "They know a lot about what they're selling, or the majority of them do. They're really big reptile enthusiasts." But most keep their animals in deplorable conditions. "I've seen turtles, reasonably large ones, in aquariums just maybe an inch bigger than the size of their body. It's not big enough for them to turn around." In order to gain ongoing access to the shops and explain away his questions about available species and prices, Bond passes himself off as a wealthy businessman who collects turtles. "I buy food or water bowls or vitamins or something. It gives me an excuse to get into the shop." In his guise as an exotic turtle fancier he's found ploughshare turtles on the block in Jakarta for around $30,000. "You can buy a reasonably nice house here in Indonesia by selling one of the ploughshares."

"Wildlife dealers are running circles around everyone," reports Chris Shepherd, another of TRAFFIC's former regional directors for Southeast Asia. "It's a joke." NGOs working to protect wildlife find rampant fraud in the Indonesian pet trade. The millions of animals exported legally each year because they are labeled as farm-raised and captive-bred are, in fact, wild. "I've been to almost every reptile farm in Indonesia, and none have breeding facilities."[5]

The creeping taxi ride back to the hotel in packed traffic is dystopian. It takes an hour to traverse one excruciating half mile tantalizingly near the hotel. No sidewalks and a thunder-and-lightning downpour make walking nonsensical.

I leave Jakarta impressed by the good nature of Indonesians jammed into their sprawling metropolis. On the short flight to Bangkok I take a crash course in survival Thai from my seatmate in row 31. She teaches me how to say "hello," "thank you," and, of course, "turtle." That takes care of the basics. And the word for "turtle"? She transliterates it from the Thai and writes the word phonetically in my notebook: *tao*. What a neat assignment as I stop off in Thailand en route to China: learning the Tao of *tao*.

— *Buddhist Turtle Credits* —

My translator's nickname is Boss. We're rolling across a steaming, smoggy Bangkok heading for the Chatuchak Market, notorious for its animal stalls jammed with dogs and birds, frogs and snakes, and—no surprise—turtles. "They scratch your fingers and it tickles," Boss is telling his turtle story. A Buddhist, he's engaged in turtle catch-and-release rituals. Believers buy turtles and then set them free in order to earn "merits"—credits for their supposedly good turtle deed that translate to things like longevity, fertility, and wealth. Shops that sell the turtles claim the animals are rescued from dangerous and unhealthy urban-sprawl habitats. "It's an attempt to justify an inhumane act," yet Boss followed the chanting instructions that came with the turtles he set free. A law student, he struggles to marry his rational and spiritual sides. So do his friends. "It's a very common practice among the bourgeoisie," he says about Buddhist turtle releases and his family's traditions.

The animal department at Chatuchak is a warren of slapdash quasi-open-air shops. Puppies stuffed into a box in one. Flocks of caged birds in another. Hot, still air perfumed with the stink

of captured critters. And turtles. There are lots of African sulcata tortoises (*Centrochelys sulcata*), a Fly River pig-nose (*Caret-
tochelys insculpta*) for $75, and an alligator snapper for $100
(bred in China, asserts the peddler). One place adorns its glass
aquariums with a handwritten demand: "If you don't buy no
photo. Only see. Don't knock." Next door the scene appears
friendlier. The smiling woman in charge sports a T-shirt showing James Dean's anguished face in black and white and a thick
gold chain around her neck. A hulking guy with the capacity to
look mean joins us when I ask about a leopard tortoise (*Stigmo-
chelys pardalis*) for sale. A flurry of rapid Thai follows, and Boss
explains: "CITES inspectors come to Chatuchak undercover,
and you look like one of them." CITES is the acronym for the
Convention on International Trade in Endangered Species of
Wild Fauna and Flora, the landmark agreement among governments designed to thwart trade in endangered and threatened species. My camera and notebook are making them wary.
But their business is legit, they say, with licenses for what they
sell. I produce my university business card for ID, and the
mood changes to cordial. "Thai people believe that if you
raise a turtle it will increase your longevity," Korbkaew Kaew-
udom says when I ask what motivates her local customers. And
foreigner buyers? "Vietnamese and Chinese people buy turtles
for decoration and because they believe turtles bring wealth."
Kaew-udom offers me a Pepsi—and a clue. "The stalls with NO
PHOTO signs are illegal."

This is a woman who likes her reptiles. She takes a long ball
python out of a case and cradles it to her bosom, sweet-talking
it, kissing its mouth. "Some turtles are indifferent," she points
at her dole of sulcata, "some follow you when they want to eat,
some will always be inside their shell—vigilant." More kisses for
the python. An endless parade of the curious pass on the alley,
most slowing to check out her turtles. None query her about the
provenance of her stock. "Watch out," says her muscleman, warning that *rum-yum* could be my fate if I ask too many questions at

Chatuchak Market stalls trading protected animals. *Rum-yum*, Thai slang for "beaten up."

Back in the car, Boss continues his lesson. "Releasing the turtle is a symbol that you are releasing bad luck and sins and sadness." About half an hour out of the city, the golden filigrees decorating the soaring roof of the Wat Phosop Phoncharoen provide a glittering relief from the sprawl of squat concrete storefronts and apartments. This temple is famed for its turtle pond, an artificial lake jammed with really well fed turtles. In a booth at the shore an antique couple keep busy cutting up cucumbers and filling plastic dishes with the bite-sized pieces. He's wearing a dapper thin-brimmed black fedora, and on her head is an oversized Oakland Raiders cap. A sign hanging between them advertises the cucumbers in English as a "hot item," only 10 baht per bowlful.

Which I buy, and join a young couple down at the water's edge. The cukes come with serving sticks, and we're kept busy offering them to voracious turtles at least the size of footballs. Their gaping mouths lunge at the offerings over and over and over again while other satiated turtles bask under the scorching sun. Today is the woman's birthday and they're celebrating it by collecting merits. "Any type of feeding creates merits," says the man, "merits for prosperity, peace and wealth." He believes more merits are acquired feeding animals than people, and in addition to turtles, he provides food for eels and frogs in exchange for merits. The birthday girl joins us, feeding herself flavored shaved ice from a food cart. "The turtle is a long-living animal," she says, "so if you feed turtles you live longer." They both wear black T-shirts, his displaying a classic motorcycle, hers commemorating the late and venerated Thai king. "Turtles are slow," she says, "so if you feed them, you will be prudent and determined."

Under a canopy that shields him from the still-scorching sun, Thassawadee Jakkawaro keeps watch on the turtle pond in his bright orange robe. The Buddhist monk serves as the temple's turtle pond supervisor. "If you let a turtle go, you will feel personal relief," he explains, "and personal relief is the key for getting merits." Relief from? "Anything. You cut the ties with the turtle and you get relief." Even if the turtle is caught by a turtle wrangler and sold again to another adherent, "you get relief because you're finished with it." Merits from different animals lead to different objectives. What is special about the turtle is its long life, so merits from feeding or releasing a turtle can be a path toward longevity. "You're not stealing years from the turtle because you extend the life of the turtle by releasing it." If you buy and release a turtle that's about to be butchered, the merits multiply. Location is important. A release into the wild is better than at the temple (provided the habitat is correct to support the turtle). Jakkawaro is patient (and prudent and determined) with the onslaught of questions. He seems to enjoy providing a turtle merit lesson.

It's no surprise that, out of the probable thousand turtles in the pond, the monk enjoys friendship with a favorite. Although he acknowledges that his pal can't rationally know him, he feels a relationship. After the crowds leave the temple grounds for the day and Jakkawaro sits at the pond in the evening, that special turtle comes out of the water to greet him. "He's the *japatow*— the mafia boss." Periodically the turtle crowd is culled. Turtles are relocated to a sanctuary so that the temple population doesn't explode (eggs keep hatching). "I must ask the *japatow* for permission to take the turtles." Once he failed to ask and, he tells me, all the turtles stayed out of sight underwater.

Half asleep the next morning I cross security and passport control at Bangkok's Suvarnabhumi Airport looking for the frequent flyer lounge when I'm confronted by a huge sculpture, alive with vibrant pastel colors. It dominates the otherwise Cartier- and Hermès-filled generic lobby. It's a three-headed serpent poised on—yes!—a giant turtle, the turtle representing an incarnation of Vishnu and the sculpture telling the tale of the Churning of the Milk Ocean. The Milk Ocean holds, according to a plaque at the base of the sculpture, the nectar of immortality, and the airport commissioned the piece to offer a sense of stability to passengers. It's a nice see-you-later from Vishnu as turtle.

— *Good Luck from the World's Priciest Turtle* —

The Thai Airways International Airbus descends toward Kunming, cruising over an area of vast urbanization that may well be encroaching on Yunnan box turtle habitat. Massive highway cuts. Towering apartment houses. Exactly where a few *Cuora yunnanensis* were found years after the species was considered extinct is a closely guarded secret. My Yunnan Province objective is better known: the Kunming Institute of Zoology where Professor Rao Dingqi is in charge of a Yunnan box turtle breeding project designed to develop an assurance colony—assurance that the species will survive. But first it's off to prowl the Kunming animal markets, lined with more orderly stalls than the mishmash in Jakarta. There I find piles of listless red-eared sliders waiting for buyers—sliders heaped on top of one another, jammed together in water-filled plastic tubs, motionless in dirty water, stuffed into delivery sacks dropped on the sidewalk next to rotting garbage. And not only sliders: bargain-basement turtles like palm-size alligator snapping turtles for a buck and a half.

While we're looking at cheap snappers at the old downtown Jingxing market my fixer and interpreter allows that a friend of his from Shanghai sells turtles via the internet. "My friend sells rare turtles because they cost more than these." While we're strolling, he fires up WeChat and asks his friend what's in stock. The response comes back in a minute or two. We're offered a baby Aldabra tortoise (*Aldabrachelys gigantean*), the giant from the Seychelles, as well as angonokas, aka ploughshares, the critically endangered tortoise from Madagascar. The Shanghai connection says the angonoka is the most expensive tortoise on the market; he's selling juveniles only a couple of inches in diameter for 90,000 yuan—just over $13,000. "Come to Guangzhou to pick them up if you want them," is the next WeChat message, "because they'll probably die if I ship them." What's the seller's advice for getting the illegal-to-trade animals out of China and back to the States? "If you're Chinese and you bring them to the airport, you'll go to jail," he says. "If you're a foreigner, maybe you can put them in your pants and make it out." A joking reference to convicted smuggler Kai Xu and the fifty-one turtles that he famously stuffed down his pants? Does the guy at the other end of the line really hold five critically endangered baby ploughshares at his Guangzhou warehouse? My fixer says yes. He's known the seller for several years, bought lizards from him, and exotic turtles are his specialty. Later in the day Eric Goode at the Turtle Conservancy confirms via email the WeChat pictures: Angonoka, "probably stolen from the breeding center in Madagascar."

At another Kunming animal market, another seller who specializes in exotics offers more angonokas. The price is based on size: 10,000 yuan per centimeter of endangered and protected tortoise. Despite the Great Firewall of China—the government's censorship system—the Communist Party can't manage to block and filter all things on the internet that authorities may find objectionable. A workaround by savvy buyers and sellers can be as simple as a message that says, "I want a model of an angonoka." The word "model" suggests a toy and offers traders plausible deniability if they are questioned. It also may defray official searches for keywords that without a modifier like "model" trigger further policing.

At the same time that smugglers are pushing angonokas toward me in Kunming, across the Taiwan Strait at the Taipei airport police yank a traveler from Malaysia out of line after the cops find three ploughshares in his baggage. The largest angonoka seized measures 41 centimeters long, and Taiwan customs agents estimate its retail value at about $40,000. If convicted, the trafficker faces five years in prison and a fine the same price as the tortoise.[6]

The Kunming Institute of Zoology is home to a few of the Yunnan box turtles known to exist. I'm welcomed into the cluttered office burrow of Rao Dingqi, the herpetologist credited with rediscovering *Cuora yunnanensis* nearly fifty years after the species was officially declared extinct. Soft-spoken and with an almost perpetually cherubic smile, Rao patiently tells the tale. In 2006, a mysterious turtle showed up in the Jingxing market, its handler asking its identity. No turtle purveyors knew, but a snapshot was uploaded to the internet and the crowdsourcing worked. Excited researchers, breeders, and turtle lovers learned that the species they thought was lost forever had in fact survived. Yet no one knew where. A year later a second was found in an older man's home. (He could have had it since he was a child; both the turtle and its keeper were youngsters back when the species thrived.) Rao had already been searching the wilds around Kunming for more. Of course, so had the poachers. Rao scored, finding three *Cuora yunnanensis* in a locale he keeps secret. He puts his hands behind his head as he retells the story of his success, leans back in his chair, and recalls how grand he felt that day.

"You celebrated with champagne?"

"No, beer!"

After further searching, in 2009 three more were found in a village near Kunming, kept as pets and for use in traditional medicine. By the time Rao and I meet, there are a total of ten wild-found in the assurance colony, along with ten new offspring, hatched with the help of Rao and his colleagues. There probably are more elsewhere. Rao sees them advertised for sale on the internet—10,000 yuan for babies, 100,000 yuan for adults; about $15,000 for a mature example of one of the rarest turtles in the world. Back in the States I've heard much higher prices quoted: $200,000 for an adult Yunnan box turtle and $150,000 for a close Asian relative, *Cuora zhoui*. These numbers eclipse the underground price point for most other endangered species. Or, as Eric Goode muses about the chelonian economy, "Turtles are on steroids."

The Kunming Institute is seeking funding to support its research and lobbying the government to secure the places where Rao found the three turtles and cordon them off from hunters and the public. At the same time both scientists and poachers from Hong Kong, Europe, and North America compete with Rao in the search for more survivors in the Yunnan countryside.

I thank Rao Dingqi for welcoming me into his office, where the high stacks of papers and journals, the full plastic bags tied tight, and the lizard making scratching noises near the door all contribute to an atmosphere that matches what I expect of a field scientist's quarters. Next, I search for the best words to ask an imposing last question. I want to meet one of the colony. Rao hesitates, looks pensive (the smile gone for a moment), graying hair at the temples adding to his now-official look. He fishes his phone out of his pants pocket and speaks a few quiet words. I drink some Yunnan tea from a paper cup and wait. A few minutes later a woman arrives carrying a blue plastic pail. Inside is one of the score of Yunnan box turtles known left on Earth.

Now out of the pail, she sits on the weathered linoleum office floor, expressing no interest in coming out of her shell. She's maybe six or eight inches long, half that wide, with a leathery-looking deep-brown carapace edged with a tan accent. I get down on the floor with her and urge this distant cousin of Fred to say hello. She pokes her head out just far enough to reveal her eyes and we look at each other, but when I greet her, in a flash she disappears again. I wait. She takes another peek and we check each other out again. But another noise in the room sends her back inside. "She's always shy," says Rao. Much shyer than Fred. After we commune awhile, I say good-bye and ask Rao if it's okay with him if I touch her. He encourages me, and when I give her a couple of strokes, he's smiling and laughing and reinforcing a Chinese tradition: "It's good luck!"

— *Walmart's Turtle Stock* —

From my Green Lake neighborhood hotel I head for the closest grocery store to pick up supplies. I dodge the near-silent electric motor scooters that, along with too many cars, have replaced most bicycles on Kunming streets. I skip the KFC chicken and the street stands selling something-on-a-stick. This isn't Mao's China. The grocery is a Walmart, complete with loudspeakers screaming nonstop commercial messages urging me to buy, buy, buy! Deep in its bowels, in the department with the chicken feet and the ready-to-eat ducks with heads still attached, and next to the frogs yet to croak, are the live turtles. Four what look like Chinese softshell turtles (*Pelodiscus sinensis*) languish, their long necks patiently waiting for the inevitable knife.

That evening, sitting on a balcony at the Shiping Huiguan restaurant, a musician softly serenading diners with her *zheng* in the courtyard below, I study the menu, looking for vegetarian options in a land so famous for meat eating that a local joke suggests diners "eat anything with four legs except the table." Under a picture of what looks like a generic stew is the English translation: wild turtle. I query the waiter. Yes, it's softshell. I order cabbage and eggplant, potatoes and corn, peanuts and beans.

Before boarding the new bullet train from Kunming to the Hong Kong border a visit to the Smiling Buddha at the Pan Long Temple is in order. The Buddha sits behind glass, smiling, of course. My guide says his message is simple: "Forget troubles. Everything is happy. Everything is okay. You don't have to worry because he has a belly and you can put everything in it." Comforting words, and a send-off punctuated by a barrage of firecrackers, the percussive offerings of other pilgrims.

— *Turtle Max* —

Security cameras and guards, steel cages with alarms and double- and triple-locked doors. Paul Crow calls his workplace a turtle prison. I'm up in the humid hills of Hong Kong's New Territories at Kadoorie Farm, where Crow works as senior conservation officer. We're in a lockdown for *gum cheen gwaii*, golden coin turtles. It's quiet in the bunker except for the steady gurgle of pumps circulating water into the cellblock of aquariums. We're looking at captive-bred specimens of one of the most critically endangered, sought after, and pricey turtles on Earth. The security is necessary because traffickers know Kadoorie Farm is breeding golden coins. Guards and cameras and locks were installed in 2005 after burglars hit the facility and stole turtles. The ongoing threat is real. Just a couple of years before my Hong Kong turtle trip, an armed gang broke into a New Territories home, smacked the owner in the head, tied him up, and stole the dozen golden coins he was keeping as pets.[7]

"It's just showing off your wealth and your power and your status," Crow says, trying to explain what motivates such violence and the golden coin trade. But the turtles also can be a financial investment. "In China," he says about mainland China, "they trade this species like a stock or a precious metal. Guys buy and sell stock in them and not the actual animal." Those investment animals are kept under lock and key at the farms where they're bred and raised. The farmers limit how many of the turtles they release into the marketplace, thus keeping the price inflated. "They're not daft guys," notes Crow about the traders. He's talking turtles as investment property while he's checking on his own charges, making sure all are accounted for and that they're clean and healthy, fed and watered.

Farmed golden coins can be sold legally, a reality that makes it relatively easy for traders to launder poached wild specimens. "Go down to Tung Choi Street," says Crow, referring to the Kowloon strip also known as Gold Fish Street, lined with tattoo parlors and hourly-rate hotels. "I guarantee they'll be several *gum cheen gwaii* on sale any given day." Later that day I follow his suggestion and find sidewalks jammed with shoppers packed into fish and reptile shops, some with show windows warning NO PHOTO. The shops hold permits for farmed golden coins "but if a poacher manages to find a wild one, it's easy for him to slip it to the shop and for the shop to sell it," says Crow. Who, except for a trained turtle handler like Crow or a learned collector, could differentiate between the farmed and the poached? "People who know turtles know what a captive-bred one looks like and what a wild one looks like."

Shells of captive-bred specimens take a form different from their wild distant relatives. "You look at a captive-bred one and the shell can be really ugly and gnarly, with lots of ridges in them," Crow says. Later on the Kadoorie tour he picks up a golden coin he'd trapped in the Hong Kong wilderness and admires the smooth carapace. "It's a little sad to take her from the wild because she was obviously happy out there. But at the time there were poacher traps all over the place, and at least here she's contributing to conservation." He's holding over ten grand worth of turtle, a wild turtle that's breeding in captivity, varying the gene pool of the assurance colony.

In the wilds of Hong Kong, far from the crowds and skyscrapers, now and again a trapper finds wild golden coins high in the hills where the streams they call home cascade through all-but-impenetrable underbrush. If he wants to peddle them he can find a buyer fast on WeChat or other hard-to-trace message services. "Unfortunately turtles being what they are, the way they're designed, they're so tough, so hardy, it makes them an ideal package to smuggle." Trained as a zoologist and raised in Hong Kong, Crow speaks fast and dispenses depressing news in the matter-of-fact manner of a scientist. "They already come canned."

The species is abused just because of its rarity and expense. "You can see photos of a whole golden coin being barbecued because the some people want to announce, 'Look how rich and powerful I am. I can barbecue turtles worth thousands of dollars.' It's horrific," laments Crow, before we leave the golden coin bunker and he checks the locks and alarm. "They're getting hammered from every angle." Paul Crow and his Kadoorie colleagues hope the assurance colonies they're raising one day are let loose in the wild, their numbers great enough to outperform poachers.

My flight back to the States leaves Asia the same day Malaysian customs agents, acting on what they called "a public tip-off," seized five crates labeled STONES at the Kuala Lumpur airport. The shipment came from Madagascar, and it was packed with 330 ploughshare and radiated tortoises.[8]

· **Slow and Easy Fred** ·

"Take it slow and easy," sang Dave Van Ronk, "if you wanna get along with me." As I watch Fred's languid moves around his house the song comes to mind. Fred's middle name should be Languid. Other than the rare times he acts hurried (such as grabbing at a worm wriggling from his attacking beak), Fred moves slow and easy. I watch him stretch out his leathery neck at a measured pace, a dignified move that would impress a cool cat in a beret at a West Village coffeehouse in the late fifties. Fred watches and waits; he acts Beat. But what goes on in his mind (I'm giving him the benefit of the doubt and allowing the definition of "mind" to extend into Fred's head), I cannot imagine.

— *Turtle Farming Stateside* —

Trade in turtles sporting a carapace smaller than four inches across has been illegal in the United States since 1975. The ban was established to slow the spread of salmonella. Salmonella can cause not just nasty illness; it can kill. Salmonella bacteria are found in turtle feces, on turtle shells, on turtle skin, and in the waters where turtles swim. The four-inch rule was designed to keep turtles out of the hands and mouths of salmonella-susceptible children since the cute little fellows are so seductive. "Pick me up!" they seem to call out. "Let me crawl around in the cupped palms of your hands." Children did just that, in pet shops and at county fairs. Despite the ban, they still do. In 2013, a carnie at the California Mid-State Fair in San Luis Obispo was arrested for giving away little red-eared sliders to kiddies who managed to throw Ping-Pong balls into floating hoops.[9] Two years before, turtles were prizes at the Maury County Fair in Tennessee.[10] And in 2015, turtles—and rabbits—were offered as prizes at the Cleveland County Fair in Norman, Oklahoma.[11] Turtles and rabbits! The promoters could have staged a race instead, named it after Aesop, and sold tickets.

When I was casting about, looking for a Louisiana turtle farm to visit, Concordia enticed me with its American-flag-waving website. "Welcome to the World's Largest Producer of Salmonella-Free Farm Raised Turtles," called out the site. Located in the appropriately named town of Wildsville, Concordia is east of Alexandria and almost at the Mississippi state line. I was intrigued by their detailed deconstruction of turtle law as it relates to the Food and Drug Administration ban on trade in bite-sized turtles.[12] Little kids do like sticking things in their mouths, squirming turtles included.[13] The year the ban went into effect the FDA estimated over a quarter of a million salmonella poisonings occurred in the United States, some severe and some deadly.

"The facts are very simple and plain," Concordia tells the public:

> It is ILLEGAL, not only to offer for sale, but to offer for ANY public distribution, turtles under a 4″ carapace (shell) size or turtle eggs. Whether it be under the guise of "adoption" *(i.e. the turtle is free, you only pay for supplies or shipping and handling—that is illegal pub-*

lic distribution), art, science, gifts, exhibition or the like, it IS illegal. Unless you, individually, are a bona fide scientific, educational, or exhibitional entity, such as a research biologist, exhibiting zoo, museum or school, that can provide a sworn affidavit stating such, you are not legally allowed to receive the banned baby turtles or turtle eggs.[14]

By all indicators, Concordia was not trying to find new customers for its own turtles (though it certainly wasn't trying to dissuade the pet trade):

> You can get a legal turtle that is only a couple of inches larger (and very legal) from reputable pet stores likely to be in your own area. . . . If you can't locate one, then contact us. If we can't put you in touch with a legitimate supplier of legal turtles, and you still can't find one then we'll provide you with a turtle ourselves. We are not interested in selling to the retail industry in general, we are primarily wholesalers of large quantities of turtles. We do not, however, want the public to be punished by being misled nor the industry to be harmed by this continued disregard for laws.[15]

Nevertheless, there is continued disregard for the four-inch law—as carnival prizes, for sale in pop-up pet stores, traded at reptile shows.

— *Turtle Bycatch* —

Before heading across Cajun country to Wildsville, I stopped off in Texas, at the National Oceanic and Atmospheric Administration's Sea Turtle Facility in Galveston, Texas. There, at the old Fort Crockett (named after Davey, of course, who was killed a few hundred miles west fighting Mexicans at their Alamo), and just a couple of blocks from the beaches along the Gulf of Mexico, scientists working with the National Marine Fisheries Service research and practice options for keeping sea turtle populations viable. Loggerhead hatchlings from Florida fill buckets in an anonymous building shadowed by an adjacent tourist hotel tower—rows and rows of solitary loggerheads in their sterile containers, just one to a bucket. These captive-raised turtles are used for a variety of studies—from the dangers of water pollution on turtle health, to the effects of human-caused noise in the ocean on turtle hearing, to the efficiency and design of turtle excluder devices.

Sea turtles suffer substantive losses as bycatch. Trawl nets pulled by fishing boats capture whatever is in their path that's bigger that the net holes. Turtle excluder devices—known in the turtle protection and fishing business by the acronym TED—provide an escape route for sea turtles. Steel bars running at a gentle angle from the bottom of the trawl net to the top guide turtles to a flap in the net and an automatic exit. Shrimpers collect their catch without worrying about killing turtles (or turtles crushing their shrimp), and with the turtles out of their nets their boats suffer less drag and better knots per gallon. The day I visit, docents are encouraging schoolchildren to try TEDs themselves, sending the poor kids through nets under the scorching Texas summer midday sun.

From an old Galveston hand I learn a Texas turtle aphorism: "That's so hard," goes the saying about something easy, "it's like shooting turtles off a log." In fact, he tells me, shooting turtles off a log is not easy. "They hear you coming and jump into the water."

A keen sign welcomes me hours later at my overnight stop in Alexandria, Louisiana, in front of the Redeemer Lutheran Church: "We're not Dairy Queen but we make great Sundays." I stop at the zoo a few blocks from the not-Dairy-Queen to check on the tortoise enclosure. The zoo promotes "one of its most famous residents," Big Al, an Aldabra giant tortoise. A flock of pink flamingos, looking like so much Las Vegas front-lawn décor, preen and parade in Big Al's enclosure while he and his tortoise pals laze about in their batten-and-board hut. Zoos depress me almost as much as pet shops, but Big Al's quarters look decent enough—for captivity.

— *Farming Turtles on the Bayou* —

The next morning I race the rented car past cornfields toward Jonesville and across the Black River to unincorporated Wildsville. I spot the landmark sheriff's substation and the beauty shop I was told to watch out for so that I don't miss the left turn off Route 84 that brings me to the Concordia Turtle Farm. Founder Jesse Evans greets me in the icy-cool air-conditioned farm office, where he's looking relaxed, sitting deep in an office chair. He quickly proves to be a man happy to spin his founder's yarn one more time. There's not much hair left on his head and he makes mention of the heart trouble that's slowed him down, but advancing age and failing health do not diminish his relish for telling his rags-to-riches tale.

"In 1968, I was just lookin' for somethin' ta do," he starts at the beginning, his soft Louisiana drawl suggesting he is in no hurry to get to the punch line. "When I was ten, twelve years old I always did catch baby turtles out in the river. We was raised poor as anyone's ever been raised. But the other kids, they thought I was rich because I always had spendin' money in my pocket. I was makin' my own money, catchin' baby turtles and sellin' 'em." Those schooldays were back in the late 1940s and early 1950s. Local rivers (Jonesville is at the confluence of four rivers) and lakes were fat with turtles, easy pickings for young Jesse's already practiced hands. "I just had a love for 'em and for catchin' baby turtles. I was good at it, too. There wasn't no doubt about it."

Turtles get out of the water and bask on limbs and logs, Jesse says, happy to explain their habits and share his snaring technique. Little Jesse second-guessed where turtles would jump when startled. Startle them he would. "I had a little dip net and I'd run that little dip net down there and there he'd be." The old Jesse smiles at the memory. "Sometimes I'd get two or three in a dip 'cuz I didn't know they's all down there." The local five-and-ten-cent stores sold baby turtles back in those good old days before the FDA rules shut down the legal marketplace. Baby turtles brought him ten or twelve cents each, he remembers, at a time when the money he earned from one turtle could buy a bottle of pop, a handful of cookies, or half a gallon of gasoline.

Concordia is not the first commercial Louisiana turtle farm. Jesse Evans saw the success of a few others and decided to farm what as a kid made him pocket money catching in the wild. "I worked offshore for about twenty years. I didn't care nothin' about that. I was a captain on a boat haulin' people back and forth to the drillin' rigs. That was the worst job I ever had in my life." He finally quit and started raising turtles with his wife, Avis. "For the first two years we broke even. We didn't make no money." After a few years the turtle farm started to prosper. "We weren't gettin' rich," he tells me, but they were paying their bills and turning a profit. He credits Avis with the decision to look at opportunities overseas and start exporting farm-raised Louisiana turtles to Asia.

Boom times for Concordia were the late 1990s and the early 2000s.

"You mean you were making some good money on turtles," I suggest.

Another broad smile from Jesse as he agrees, "I was making some money! The best year I have had was in 2002 and was $3.2 million." We're sitting in rural Louisiana, out past the cornfields and the sheriff's substation and the beauty parlor, on a farm that turns turtles into cash. "That's a lot of turtles," I tell the seasoned entrepreneur.

"That's a lot of turtles," he agrees, "and that's a lot of money for me, I can tell you that." He laughs and so do I, at the oddity of making a fortune raising turtles. "I enjoyed what I was doin'. It just thrilled me to go out every day, pick up the eggs, come in and clean 'em up, put 'em in boxes. Then in sixty days they would be little turtles."

"It's not just thrilling to your pocketbook?" I ask, and he insists that the money was nice but it was the magic that thrilled him.

"It didn't excite me to make lots of money. I was paying my bills," and Jesse says he was glad the bills were paid, but "beautiful" is the word he uses as a prelude to describing that miracle "when they come out and you clean 'em out of the egg shells and put 'em in a big pan—a thousand in a pan—and you look at 'em and say, 'Man, ain't that somethin' to look at those little guys, and they're just workin' and crawlin'. And I," he says about his midwifery, "done that myself."

"You and maybe with some help from somebody upstairs," I suggest, pointing heavenward.

Another smile as he says, "That's exactly right—a lot of help from the one upstairs."

Along with help from the turtles. When the turtles are past their breeding prime, Concordia could sell them for meat, but Jesse and his son Davey—who now runs the farm—instead put them out to the equivalent of pasture. "Davey said, 'Daddy, I don't want to sell my turtles for somebody to eat. They've been too good to us,'" Jesse tells me with parental pride. So Concordia uses some territory along the Black River—sloughs and little lakes—where the old turtles live out their lives in a turtle paradise. "They go where they want to and do what they want to do."

Not that Jesse and the Evans family shy from eating turtle. "We eat the softshell turtles, the snappin' turtles and the alligator snappin' turtles. They delicious." Another smile. "In fact, that's what I'm havin' for my lunch today, turtle . . ." And then he says something I simply cannot understand. I ask him to repeat it and I still cannot wade through his accent. "That's French," he says, trying to help me. We laugh at my inability to figure out what he's saying, and I ask him to write it on my notepad. It's "turtle sauce piquante." I finally understand. The turtles the family farms would taste good too, if the family used them for food—yellow-eared sliders from Georgia and the Carolinas, and red-eared sliders and river cooters native to Louisiana. But the farmed turtles do not end up in Avis' iron pot.

Louisiana softshells can grow to twenty-five pounds, Jesse says, "About this big," he says, making a shape with his hands the size of a basketball. That's a turtle that costs about a dollar a pound when it's sold door-to-door by locals to locals and yields as much as eighteen pounds of meat. "Once you eat one of 'em, you eat another one," he says as a testament to the softshell taste. "It's good." And it doesn't taste like something else. "It's got a taste of its own." Not like chicken? "No, it don't taste like chicken." Avis, he says, southern-fries it. "She flours it, like you do chicken. She puts it in an iron pot. She puts a lid over it. She turns the burner down. And it'll just cook and steam. When it's brown on one side, she'll take it and turn it over, and let it do the same thing. And that steam," he's appreciating what it's doing to his lunch, "you just suck the meat off the bone."

— *Turtle for Lunch* —

I wangle an invitation to the kitchen. "Let's go," Jesse says. He's enthusiastic and hospitable. "Let's go see if she's got it cooked." We walk from the office over to their home—a sprawling place that would look appropriate gracing a golf course. Avis shows off both the stewing turtle bones and a pan of fried green tomatoes. Her thick white hair is brushed back from a face that seems just barely to tolerate my invasion of her kitchen, though she is quick to offer me a taste of her tomatoes. I wonder what I'm going to do if she offers me some turtle bones to suck on.

While the turtle steams, Avis and Jesse regale me with a menu of what else graces their dinner plates: deer, rabbit, squirrel, and the bounty of their garden.

"Have you noticed the shape of ninety percent of the American people?" Jesse muses, figuring the cause is that those who become obese not only eat too much but also eat the wrong food. "That what you buy in the market is so full of preservatives, you don't know what you're eatin'." We talk about the feedlots and hormones and antibiotics of the corporate animal farms, and about what a typical American diet likely does to a typical American's health. "They done got lazy, too," Jesse notes, while Avis adds with a caustic smile, "The government's made it too easy while people like us work and pay taxes." Hmm. They laugh and she amends her remark with an acknowledgement that "some of 'em work hard.

"Did you like that tomato?" She changes the subject, and I compliment the crispy breading and the surprise of the sweet, juicy inside. "See, this turtle is coming off the bone?" She lifts the lid on the skillet. "I usually put it in a pressure pot, but it cooks so much better in this iron pot." Jesse looks around the bounty in his kitchen and announces, "They'll never starve us to death here in Louisiana. If they have a famine to hit this country . . ." Avis interrupts with, "If they take the guns away . . ." But it's Jesse's turn to interrupt. "We can still set traps and snares. If we have to, we'll make a hook out a straight pin and catch a fish on that." He's smiling and laughing. "We might need a little salt and pepper. But they got a salt mine about fifty miles from here, and we can go up there and scratch enough salt to last us a while." The talk strays to city folk who think food comes from a supermarket and don't know how to hunt or grow their own—and we're interrupted: it's time for my tour of the farm with their son Davey. I'm spared deciding if I'll try the simmering turtle (or perhaps Avis never considered inviting me to lunch).

— China Learns a Cajun Lesson —

I find their son Davey in Concordia's hatchery, across the drive-way from the office. This is where the turtle eggs wait to hatch. Collected eggs are washed and soaked in Clorox-infused water, and they're treated with an antibacterial—all in an effort to keep the eggs and hatchlings free of salmonella. Once out of the shell, a representative sample of every twenty thousand little turtles (Concordia deals in vast numbers) is inspected by a Louisiana State University vet who certifies the batch salmonella-free. But keeping turtles from spreading the bacterium is a chore that must continue after they leave Davey's custody.

"Lemme tell ya somethin' about a turtle," he says, and I am ready for a lesson from a guy who's lived his life among them. To stop the spread of salmonella, he instructs in the same soft Louisiana lilt as his father's, "Don't worry about the turtles, worry about yourself. After you handle a turtle or play with a turtle—kids, grownups—wash your hands. It's a sanitary deal." Indeed, after he showed me some hatching eggs, first thing Davey did was wash his own hands. "You can lock a dog up in a pen"—the word is stretched out: daaawg—"and he'll walk around in his feces. You go play with him and the dog licks you in the face. Well, he's gonna lick his feet and his . . . ," he avoids specifying the other body parts dogs tend to lick. "Well, you're gonna get salmonella. I don't care what kinda animal it is." No matter the clean bill of health his turtles receive when they leave Concordia, it's the job of the next custodian to keep them and their environment clean—and to wash up.

"They're raisin' their own turtles now," he tells me about the Asian marketplace. It's a humid 92 degrees in the hatchery, just perfect for the fledglings. Davey is wearing his work clothes: blue jeans and a well-worn green T-shirt decorated with a few holes. The boom days for exporting turtles to China are history, he says. Not that the appetite over there is satiated, but why there is such a lust for turtles in Asia remains a mystery for him.

"I have no idea," says the man who's shipped literally millions of turtles across the Pacific. "All I know is that the red-eared sliders, when we first started sendin' them, were goin' for food." And for breeding stock, he learned, so that Asian entrepreneurs could start raising them for the Asian market. "When they got everything they wanted big enough to breed, our sales collapsed. That market is gone."

"You supplied the turtles that put yourself out of the China business?"

"That's right," is his soft reply. "We made lotsa money doin' it," he recalls, marking the end of Concordia boom years—when a million and a half turtles a year left Wildsville for Asia—as around 2005. But just as wild salmon commands a much heftier price at the Whole Foods down the street from my university than its farm-raised cousins, wild North American turtles sell at top dollar to Asian customers. "A lot of turtles are wild-caught here and sold to New York, Chicago, . . ." and he lists Chinatowns from coast to coast. It's still a good market, he says, even if it's not always a legal market. "Some guys, it don't matter to them if they can get away with it."

The bulk of Concordia's business is Stateside now, selling four-inch-plus turtles—that minimum legal size—into the pet market; the family-owned company's annual yield doesn't exceed three hundred thousand turtles. Most of those turtles go to Petco and its almost fifteen hundred big-box pet stores in the United States and Mexico.[16]

"Quite the collapse," I say, and he lets out a pained hearty belly laugh before saying, "You can't blame them," about his Chinese competitors. "Back in the heyday of turtlin', I used to go to the airport in New Orleans twice, three times a week with a hundred thousand." When Davey laughs, his sun-weathered face often displays an intent, concerned look that's reinforced by his short, mostly white hair. "That's the American way. I don't blame nobody for that." Concordia remains in the black, and Davey says he loves raising turtles. "We're all still gettin' a paycheck

every week. We're eatin'." I note that his mother is cooking turtle while we're talking with each other. "That's right," his appetite shows with a smile. "I love it," he says about eating mom's
turtle. But he's just as enthusiastic about turtles that do not end
up in her skillet.

"A turtle is a great pet!"

Why?

"You can have turtles where you can't have nothin' else." He
keeps smiling, happy to talk about—as Simon and Garfunkel
sang—what keeps his customers satisfied, pointing out that a
pet turtle can be happy living in places unsuitable for other
animals—like an apartment. "You can talk to a turtle. A turtle
understands you when you talk to 'em. I got two or three—I'm
probably gonna let 'em go—I call 'em and they crawl right up
the bank." These are old red-eared slider breeders about ready
to be let loose, turtles long in Davey's care. "I had one called
Sally, she'd eat right out of your hand."

"You're not really talking to them like you talk to your dog,"
I argue. "You can't tell them to bring you your slippers and newspaper."

Davey rejects my dismissal of his turtle talk just because they
don't fetch. "You can say, 'Come here, come on,' and Sally'd
come right on up the hill—just like you'd call your dog to come
to you." I keep arguing the point, questioning if he could see
in Sally's eyes the kind of understanding that's evident when
he talks to his dogs. "That one turtle you could," he says about
her. "She was just a gentle ole turtle. She was so old that her
red had done turned to green." We leave the relative comfort
of the hatchery and amble into the suffocating Louisiana
summer heat

"She's coming because you've got food for her," I counter, suggesting he and Sally are not necessarily pals talking. She's just
learned that his voice—his presence—equals feeding time. But
Davey insists there's more to his turtle relationships. We're walking through the midday heat past ponds filled with breeding

turtles. "You see these turtles in these ponds? I could have sold these turtles to a Californian. For meat. To be butchered." He gestures to his family's holdings, pointing to their stately home, the outbuildings and farm equipment. "Everything you lookin' at here: houses, trucks, cars, kid's cars, kid's houses, come from those turtles. "I'm not going to cut them turtles' heads off." And he proceeds to explain in detail about Concordia's turtle old folks home. "You know what I'm doin' with those turtles? I'm catchin' 'em out. I got three big lakes on my property that those turtles bought. I'm taking 'em down there. There's plenty of hydrilla and duck grass—stuff they like to eat. I'm lettin' 'em go in the lakes. I'm not gonna kill 'em. I'm gonna let 'em go." It's impossible not to like this guy. He's happy with his work, he doesn't lust after more money, and he loves his turtles. "Everything I've got comes from them. Everything!" His melodic Louisiana accent adds compelling charm to the story. "I'm not punishing my turtles for being good to me. So I'm gonna let 'em go."

Davey Evans walks to the water's edge at one of the concrete pools and dangles a piece of foliage. Brave fat red-eared sliders come up out of the water to check him out. I look across the pond at bobbing heads, too many for me to count. Ten thousand breeders, Davey tells me. It's quiet except for a chorus of songbirds—peaceful country, far from the strip malls that house Petco. "These animals are content," Davey says. "They're happy."

How can we determine if an animal is happy? Especially a turtle. Is Davey happy? He's smiling, apparently easygoing and enjoying telling me his life story and Concordia's story. Am I happy? I like to think so, most of the time. I like Davey's answer about his turtles' happiness quotient. "They seem to be happy. They get all they want to eat. I really don't know if they're happy or not. They just seem to be happy. I know what: I've had that openin' in the fence," he points to an escape route, "and they just don't want to leave."

As for Davey, he claims I'm correct. "Oh, I'm a happy man.

And she don't look sad," he says about a turtle that's chasing the plant he's dangling in front of it.

Next stop is a spit of sand next to the ponds, sand where the turtles lay Concordia's bounty. A few workers are digging in the sand for the fresh eggs—eggs a little smaller than chicken eggs—carefully pulling them out of the ground and lining them up shell-to-shell in metal carrying baskets. This is hot, sweaty, labor-intensive work.

The turtle trade—legal and illegal—is packed with wild-caught turtles. But Davey is convinced his farmed product makes for a happier turtle than one captured out of its riparian homeland. "That would be like somebody catching you at your apartment or your home," says Davey, storyteller that he is, "taking you to a strange place and just lettin' you go in it. You're gonna just crawl and look and hunt to get back to your house. You're gonna get out!" Contrast that, he says, with his turtles. The wild is an unknown to them. They hatch. They're moved to a tank where they grow until they measure the four inches that makes them legal for the U.S. pet market. "They don't know nothin' but captivity." Hence they are a better pet, he says, because they were raised captive.

"There's enough going on in their little turtle brains," I ask him, "for the captive ones to feel at home and the wild ones to want to escape?"

"I hunt a lot," he's answering with a story, "and I've never found nothin' that wanted to die or get trapped. You know what I'm sayin'?" With that example, the country boy and the city boy understand each other. But he's not done. "Fer instance, you. You want somebody catchin' you and puttin' you somewhere."

"No, thank you," I say.

"You wantin' to go home, know what I'm sayin'?"

"I know exactly what you're saying."

"My turtles, they know nothin' but captivity. As long as he's gettin' fed, he don't care where he is."

I wave good-bye to Concordia's enormous bale of turtles and

retrace my tracks over the Black River to Jonesville. Hard to miss the sign announcing the site of the "second tallest Indian mound in North America." Hard to miss the sign, but it's easy to miss the mound. The sign shows off an under-construction replica. The original pre-Columbian earthworks were knocked down in 1931 for a bridge over the Black. In the 1970s archaeologists digging in the ruins found pottery remnants, human remains, and deer and turtle bones.[17] Across the street from the mound is the Auto Matt self-service car wash, adorned with a familiar cartoon of a turtle wearing a top hat. "We use Turtle Wax," proclaims a sign above the hoses. With the air conditioner and Beethoven's Fifth (courtesy of Red River Public Radio) both cranked high, I head back through the cornfields toward Alexandria, past signs offering beignets and shrimp and a gas station hawking souvenir alligator heads.

About that Turtle Wax: company lore has it that founder Ben Hirsch invented the first liquid automobile polish in his bathtub back in the 1930s and at first called it Plastone. In 1946, he renamed the stuff Turtle Wax. In 1954, goes the story, Hirsch sat down at the piano and banged out the "Turtle Wax gives a hard shell finish!" jingle.[18] The song first became popular in an animated TV spot featuring a Superman-impersonating turtle, with Jimmy Durante's voice, proclaiming, "Toitle Wax Auto Polish with Brillium gives your car a super-brilliant hard shell finish that protects it like an invisible garage!"[19]

— *Pond vs. Cage* —

It's sunrise on this late summer morning and I've parked myself on a bench overlooking the Delta Ponds—a swampy stew along the Willamette River not far from my campus office. This is reputed to be a hotbed of wild turtle activity in the midst of urban sprawl. I left my old Volvo on the edge of the Valley River Center—a generic shopping center, always depressing to me, that eviscerated downtown Eugene after it opened in 1969.[20] I stroll along the familiar bike path—usually I speed past here when I'm on my bike, anxious to get past the Macy's, the movie theater, and the often-stinky stagnant ponds. But this day I stay pondside and sit on the bench with a cup of shopping center Starbucks, the latest copy of *The New Yorker* (in case this is a long-lasting stakeout), and my binoculars. Songbirds and ducks compete with the constant hum of nearby freeway noise. I hear an approaching train whistle as I wait for the sun to warm us all: me and the rare turtles I've been told I'll find here so close to my home.

The western pond turtle (*Actinemys marmorata*) is listed by the Oregon Department of Fish and Game as "a species in need of help," primarily because of habitat loss. In Oregon it is illegal to do anything but look at these turtles—no catching, transporting, buying, or selling. These ponds with shopping center neighbors are a habitat loss exception. Once commercial sand and gravel pits, a Eugene civic project restored them to western pond turtle heaven, and now the ponds are home to salmon heading to the Pacific, a long list of bird species, and a tally of at least sixty western pond turtles.[21]

It's just after seven in the morning as I scan the ponds through the glass and watch a duck coast along the surface, poking its beak into the water. Footfalls and a snort from a jogger disturb my reverie. A siren punctuates the nearby freeway traffic hum. Voices from a chatting couple of bikers add to the soundtrack, their bike tires buzzing along the path pavement.

But no sign of turtles. Yet.

I wait.

A goose interrupts the murmur of the bird-and-traffic chorus, as does an air conditioner compressor that kicks on in a building behind me, a building I had not noticed until it interrupted my idyll with its incessant nearby motor noise. Now come a couple of joggers talking, one bragging about past runs, "Until I was, like, I'm so tired."

And still no turtles.

Maybe it's the coffee, but now I'm getting impatient. Where are the bloody turtles? Cue the turtles so I can note them as coexisting near my backyard and get on with my Saturday. Two huge herons lope like pterodactyls over the ponds, now flooded with sunshine. I get up and stuff the magazine in my back pocket and pace the path, stalking the shorelines for turtles. A snow-white egret reflects its perfect twin in the still water. Enough of this bucolic waiting. I'm going home for breakfast and figure I'll come back later in the day when maybe it's warm enough for the poikilothermic western ponds to brave the day.

After noon I'm back at the ponds, the day has warmed up to the comfortable mid-70s with a nice breeze, and as I look out on the water, there on a rotting log is a fat duck, a perched heron, and between the two—straddling a branch of the log—is a plump turtle resting and looking as if it hasn't a care in the world, its legs hanging over the log, its tail dangling out and its neck stretched, holding its head high as it looks out over its lily-covered pond domain. No doubt this is a western pond: dark brown to olive carapace with similar coloration for the legs and head; likes to bask in the sun.[22] And carefree? The adults suffer little natural predation. It's a fat and happy western pond turtle, and seeing it makes me happy.

I cross the parking lot and force myself to push open the Petco door. It's cat adoption day, and I stride past the feline temptations, past the goldfish jammed in a tank and the ferrets in a glass cage, to a terrarium labeled TESTUDO HORSFELD [sic] holding the only turtles I find in the store: two Russian tortoises, *Testudo horsfieldii*, each about the size of a Big Mac. I stop and watch. One sits inert, its head and legs deep in its shell. The other is jamming his head against a far corner of the cage while its front legs claw at the glass. This guy wants out: there is no other possible interpretation of his actions. What if I were to take him out? What would I do with him other than put him in a bigger box? His natural habitat is arid Central Asia, not

temperate rainy Oregon. Not to mention the price tag: $139.99.
I watch for several minutes as he keeps clawing at the wall, a
barrier absurdly decorated with a photomontage of the rocks
he wishes he were free to climb.

A friendly Petco clerk passes me and offers a smiling, "You
finding everything alright?" Not really. Certainly not all right.
"Docile, lifelong companion," reads the description on the terrarium. "Very social and curious, will roam and explore. Adept
at climbing and digging." Number two sticks its head out and
makes for the wall. The smiling clerk is distracted by a parakeet
that's escaped its cage and is darting about the store. "Oh no,"
says a browsing customer, "it's over by the cats." Next I hear,
"Good catch!" and the immediate Petco crisis is over. The bird
is back in its cage. The Russian tortoise is still clawing at its faux
rock wall. I refrain from telling him about the wild life his cousin
is enjoying just across the parking lot.

— *Slider vs. Pond* —

Now I'm in Springfield, just across the river from Eugene (yes,
that Springfield: home of the *Simpsons* and the place where Bart
wrote over and over on the blackboard, "I will not spin turtles.").
I'm sitting in the office of Oregon Department of Fish and Game
biologist Chris Yee, talking western pond turtles. Yee's bushy wal-
rus moustache and shaved head combine to make for a natty
complement to his quick-and-easy smile. On this drizzly October day, he looks ready to join a barbershop quartet singing Cole
Porter classics: "Is it the good turtle soup or merely the mock?"
Instead he regales me with tales of his struggle to help western
pond turtles survive amid urban sprawl, tiny state government budgets for nongame wildlife protection, and aggressive
populations of red-eared sliders. Later, when we go out into the
field, he casts a slight doubt on my western pond sighting—the

carapace of some red-ear sliders, when they age, turns dark and can easily be mistaken for a western pond.

Springtime both species come out of the water to lay their eggs on land. Females find what they consider a good place to dig their nest chambers. They urinate on the ground to loosen the soil. "It helps her form that nest chamber." The dig is a precise process, leaving a hole about five centimeters in diameter. "When you watch them do that you see that they take the hind leg, put it in that hole and scoop dirt out like the leg is a little shovel." Yee is fascinated by the neat footwork. "It's not like a dog digging a hole." Both the ponds and the red-ears dig such nests and in the same locales. They lay their eggs in the nest openings, "and then they form a plug over the top of the nest chamber with that dirt they excavated and pack it down." The red-ears hatch earlier than the ponds, so predators find vacated nests and empty shells littering the landscape. They quickly learn where to look for eggs, and after the ponds lay theirs, it's open season for the raccoons, foxes, skunks, and coyotes coming out nightly on the prowl. Unfortunately for the eggs, "they'll find the majority of the western pond turtle nests." Urine helps the egg eaters find the pond eggs. "The red-eared sliders, from what I've seen, as soon as it warms up in the morning, they're on land laying their eggs. The nest plug has urine on it and it has the entire day to dry up and have some of that urine smell dissipate." Not so the ponds. "Some of them come out first thing in the morning," observes Yee, "but others have been seen out even after dark laying eggs." The urine remains pungent, a GPS for predators out on the prowl looking for eggs. "It's a real recipe for getting your eggs eaten," Yee says about the nighttime laying. Blame urbanization, not the western pond egg-laying technique and schedule. Predators thrive on pet food and other human discards, and shopping centers and tamed rivers force western ponds to nest in close proximity on what land they can find, making their progeny easy-to-come-by prey.

Come springtime Yee invites me to inspect a trap he's set up at a pond packed with red-eared sliders and no western ponds. Turtles coming ashore seeking nest sites find themselves directed along a fence into a corral designed to hold them until Yee collects them. The one-way gate that keeps them from escaping reminds me of the signs at Hertz car return lots warning of "severe tire damage" if you change your mind and back up. He designs and builds the red-eared slider traps himself and checks them daily during the late spring and early summer laying season—all on his personal time off from his state job.

Biologist Yee tells me he doesn't like the pond turtle more than he likes red-eared sliders, but that protecting local turtles from the non-native invasives requires him to dispatch red-ears. "Humanely euthanize" is the term of art for what sounds like a gruesome task until Yee explains why he considers it the best practice. "Because reptiles are so tough, you really have to take extra care to make sure that you are not causing the animal any kind of suffering." First, Yee injects the turtle with both a sedative and a dissociative—a drug that creates a sense of detachment (for that he uses ketamine, the so-called date-rape drug, Special K). The cocktail goes into a forelimb. "I wait until it's knocked out," Yee says in a measured, matter-of-fact voice. "I decapitate the turtle and then I pith the brain. Because reptiles are so tough, you have to pith the brain. If you don't, that head—even though it's cut off and it's not getting blood to it—remains active. We don't know how long it remains active, but it could be hours." At this point it doesn't feel to me that the story would be complete without knowing the technique he uses to pith the brain. "I destroy the entire brain, all of the tissue. I get in there and I scramble that up." With what? "I usually use a screwdriver. I dig around with it and make sure I've done the job properly."

That gruesome job brings to mind a tale collected by folklorist A. W. Eddins a century ago from a Texas turtleman working the Brazos River delta. Two boys fishing come upon a decapitated turtle, its legs still moving. They argue about whether it's dead or alive, one insisting it must be dead because its head is severed and the other countering that since its legs flail, it is alive. They take their disagreement to their uncle, who contemplates the conflict and declares, "Now, boys, it's just like this. That there turtle is dead, but he don't know it."[23]

Chris Yee would not choose to euthanize red-eared sliders. "It's the part of working with turtles that I dislike," he says, after describing the process in such florid detail. "A lot of these red-eared sliders were born here and raised here." They're classified as a non-native species, but like so many offspring of immigrants—myself included—Oregon is their home. "They are quite remarkable, and I have a lot of respect for them. That's why, when I euthanize them, I take great care to try to cause the least amount of discomfort during that whole process as I can." He collects data on the remains in order to make research use of the animal. "Because they are a problem for our native turtle, that's just part of my job."

— *Turtle Personalities* —

At a deep pond isolated on a dead-end road near the Eugene airport, I join Yee in a leaky flat-bottomed aluminum rowboat to pull turtle traps. Still water and thick shore foliage greet us. Downed trees in the water offer turtles ideal basking spots. Were it not for truck traffic on close-by Highway 99 and the periodic plane buzzing past on final approach to Mahlon Sweet Field, this could feel like isolated wilderness. Yee is after healthy Oregon western ponds to compare with western ponds from the Columbia River Gorge. Just a few hours' drive north of this pond, Washington State turtles are suffering from a mysterious disease that results in ulcerated plastrons. This day we find three specimens that we weigh and measure and keep for a CAT scan. I help by holding the control group, and see quick proof that turtles can express different personalities. The first one we grab is docile and waits patiently in my hands while Yee notes its stats and paints its carapace with hot pink fingernail polish as an easy-to-see notice that it's been cataloged. The second squirms some in my grasp. But the third is manic. He doesn't want to be restrained, and it takes much of my finger strength to keep his powerful legs pushed into his shell. Before we're finished with him his fast-flailing legs make contact with one of Yee's hands and the turtle's sharp claws draw blood. He's wild. Three western pond turtles from the same place, and three unique individuals.

— *Ninja Turtles Fuel Crisis* —

There's no question in Yee's mind that the pet and food turtle businesses are to blame for the red-eared slider invasion world-wide, and for a red-eared crisis in their home ranges. "Red-eared sliders are very easy to breed, and as a consequence they're a very popular turtle in the pet trade. There are some areas in their native range where they're in trouble because people re-move them illegally for the pet trade." The poaching, breeding, buying, selling, and discarding of red-eared sliders has been problematic since well before the Teenage Mutant Ninja Turtles craze. But the extraordinary popularity of the cartoon charac-ters perhaps exacerbated the crisis, encouraging their fans to buy pet turtles, a caretaking responsibility too many found easy to abandon. Now there's an invading North American red-eared on every continent except Antarctica.

Comics illustrator Steve Lavigne went to high school with one of the Ninja Turtles' creators, Kevin Eastman. After Eastman and Peter Laird conjured up the four crime fighters, Lavigne went to work illustrating the anthropomorphic turtles. When I contact him to talk Ninja Turtles I learn, no surprise, that his family has its own personal turtle story. "We had a turtle," he says, enthusiastic to share the fond memories. "My kids grew up with it, and of course it got named Michelangelo." (Kevin Eastman says the Michelangelo character in the comic was based on Lavigne.) "Pretty amazing," he laughs at the memories, especially about how surprised he was to learn that turtles can move fast. "And I never knew they made noises. It freaked me out one time just on the couch and hearing . . ." and Lavigne makes a soft squawking noise.

"I'm sure of that," artist Lavigne agrees when I ask him if there's a correlation between turtle pet sales and the popularity of the Ninjas. "I think every kid thought they wanted a turtle because of the Ninja Turtles." He figures kids probably expected their pets to walk and talk like the Ninjas. "I'm sure pet stores sold a ton of them." But he's sobered at the suggestion that the popularity of Ninja Turtles may have contributed to invasive species disrupting the balance of nature. "I'm a little horrified because I love turtles, but when they're in the wrong ecosystem it's not a good thing. I know an invasive species can wipe out another species. It's scary."

"It's kind of freaky that the success of the Ninja Turtles made a lot of people go out and buy pet turtles," agrees cocreator Eastman. Not that it's an artist's fault if a naïve or irresponsible parent dumps an unwanted turtle in woods.

Another example of good intentions backfiring occurred in the 1990s when Oregon chose to outlaw the red-eared trade. Biologist Yee figures some pet merchants decided it was too expensive and too much of a hassle to ship their unsold stock back to the wholesalers. "They took them out to the local ponds and rivers and dumped them." The result is a flourishing Oregon invasive red-eared population at the expense of the struggling native western ponds. Bored pet owners worldwide continue fueling the red-eared population with discards.

Add to habitat loss and red-eared invaders another assault on the western ponds: malicious mischief. Chris Yee's turtle fieldwork includes encounters with difficult-to-fathom brutality. Kids playing in ponds upend turtles and spin them on their backs; they stack them up and then knock down the stacks, "and in some cases just take a rock and smash the carapace, which is lethal." When he encounters these human animals in the field, Yee tries to pass along his own fascination and respect for turtles, hoping to counter whatever motivates such mindless abuse.

"They are a charismatic species!" Chris Yee says, refusing to yield their potential appeal to the usual poster children for endangered wildlife like pandas, seals, and chimpanzees. "Turtles are easy to love. They don't attack you, so there's not a fear of turtles." Not for Yee, maybe, but chelonaphobia does exist— there are people who are afraid of turtles. *Huffington Post* news editor Ani Vrabel publically confessed her fears back in 2013. "I've had terrible dreams about turtles flying at my face, chasing me with the intent of gumming me to death, and hiding in my bed," she wrote in a turtle-terror memoir. Vrabel made a trip to the New York Turtle and Tortoise Society's annual Turtle and Tortoise Show, where she faced her nemesis (many more than just one of them, of course) and learned to appreciate what she now calls the "awe-inspiring" animals (even though she still wants to keep her distance from the beasts).[24]

• **Fred as Valium** •

Fred, I'm realizing, slows me down. He's Valium on four legs. If I come home from the university wired, stressed, or rattled, a few minutes watching Fred, hanging out with Fred, is therapeutic. Odd as it may seem—especially for a go-go personality such as mine—I'm finding that I can sit and watch Fred, even if he barely moves, and become mesmerized. I'm rarely studying him. I'm pondering him. And that makes me ponder the rest of us. Often I find myself just watching and not even pondering much. Just relaxing, slowing down, and leaving what a journalist friend of mine calls "the dismal details of the daily downer" further and further behind. Thanks, Fred.

— *Incarcerated with Ponds* —

Two men who are not afraid of turtles—especially western pond turtles—are Washington State prison inmates Terrell Hill and Joe Goff. The two "residents," as the warden calls prisoners, live about an hour's drive across the Columbia from Portland, isolated deep in the Pacific Northwest woods at the minimum-security Larch Corrections Center. I arrive on a typically gray November morning: it's drizzling and a low fog hangs on the evergreens that surround a collection of bland look-alike one- and two-story frame buildings. The place looks as if it could be an anonymous suburban apartment complex, if the high cyclone fence topped with concertina wire were erased from the picture. A sign in the administration building notes that "problems become opportunities when the right people join together." A booming loudspeaker breaks the remote quiet, bouncing prison announcements around the valley. "Offenders Goff and Hill to the turtle facility," I hear as I'm escorted through locked doors and gates to a hut where Hill and Goff tend the turtles in their care.

Terrell Hill has spent a dozen of his thirty-four years in prison when we meet. He's a gregarious fellow with a mouthful of gold-emblazoned teeth that flash in accompaniment with his welcoming smiles. He gently grabs a western pond—a specimen so new to the prison she's yet to be named—from a tank that looks like it could double as a back-porch hot tub. He flips the turtle over and shows me the mysterious inflammations on its plastron. The brown and tan of the turtle's shell look color-coordinated with Hill's prison-issue khakis. "We take care of them and nurse them back to health to the best our ability," Hill tells me. "We're just basically taking care of them, trying to give them the best opportunity to go back to their habitat healthy as possible so they'll be able to thrive when they go back into the world." That the same could be said by a warden of prisoners is not a comparison missed by Hill and Goff as they contemplate their own circumstances.

Their efforts are part of a novel partnership between the prison and the Washington Department of Fish and Wildlife. The sick pond turtles come from the Columbia River Gorge, where close to 50 percent of the sampled pond turtle population show mild to severe signs of the disease. Ailing turtles receive initial veterinary care at the Oregon Zoo and then recover at the prison. Once healed, they are returned to the wild. Inmates volunteer for the turtle service and are chosen after a rigorous vetting process that includes an assessment of their past prison behavior, an application essay, and an interview.

"These lesions on the plastron are eating it away," Hill explains, pointing out three dime-sized sores on the turtle he's holding. While specialists labor to diagnose what is causing the affliction, Hill and Goff keep the tanks clean, feed their charges a diet that includes mealworms they raise in the prison, observe their sick charges, and wait to learn what further treatments might help the ailing turtles.

— *Turtle Therapy* —

Meanwhile, as clichéd as it may sound, the turtles are helping their caretakers. "This gives me the opportunity to care for an animal that is different. I grew up messing with pit bulls and Rottweilers and stuff like that," Hill says. Friends kept turtles, and turtles were featured in his Seattle high school biology class, but there were no turtle pets in his household. "This gives me an opportunity to try something new. I'm a city boy." But it's not just something new. Hill understands why many people find themselves, as he described it, "in awe of turtles." He figures that, at least in part, it's because of their unique ability to sequester and protect themselves in their shells.

"Awesome" is such an overused exclamation. The word is tossed into chatter with abandon that almost equals "like" and "you know." But when Terrell Hill identified people as being "in awe of turtles," it struck me that he was pinpointing the meaning of the word as defined by the dog-eared dictionary on my desk: an emotion of mingled reverence, dread, and wonder, along with fearful veneration and respect. And that litany—reverence, dread, wonder, veneration, respect—does sum up so much of the human emotion attached to turtles across cultures ever since we humans started to share space with these mysterious reptiles that wandered with the dinosaurs.

"You rarely hear them make any noises," Hill observes as he considers the who and the what of turtles. "How do they talk? How do they communicate with each other? How do they know that they're being taken care of by someone who's actually trying to look out for their well-being and not out to harm them? A lot of different factors make the turtle appealing to me." As he watches them, "stripped from their habitat," living in a plastic tub under a basking lamp instead of out in the sun and the rain, he considers what may be the positive aspects of their isolation. "In here they have no sense of being put in the position of being attacked by a predator. I don't know if that weighs on them or not." He speaks for one of his turtles: "'It's been going on two days and I haven't seen anybody trying to eat me,'" and he laughs. "I think they're pretty smart." He watches them acting aware of their surroundings. "They have their own level of consciousness. This new environment is not what they're used to, but at the same time I think that as long as God gave them eyes they can see that it's different." We're looking down into the tub of water where Hill returned the turtle with the lesions. "When they see us, they probably don't know what we're doing here with them. That's why, when I picked her up, she was squirming. She was probably thinking, 'Wow! What's going on here?'" While we're talking about her, the turtle in question is looking back up at us. I ask Hill if he thinks the turtle is relating to us as she watches us—if the turtle knows we're something different than her tub or the pipe that supplies it with water. "Yeah, yeah. Of course," he says without hesitation. "She obviously has a heartbeat and eyes that see something." What she's thinking, Hill doesn't begin to surmise, suggesting only that perhaps the turtle is worried we're potential predators. "The pipe's not grabbing her, but I am!" Another satisfied laugh.

But Hill is realistic about his relationships with the western ponds. "Naw, they don't know me." He looks at the newcomer in the tub. "It gives me a sense of purpose," he says about his caretaker role. "Something to do with my time. Something to learn. Something to get involved with. Something I cherish because as long as it's got a heartbeat, it's a life." As Hill looks forward to when he's back with his family rebuilding his life on the outside, he's convinced he won't forget the turtles. "They helped me, gave me time to reflect. This is a very important time." I tell Hill about my own experiences with Fred, how sitting with him in my living room tends to slow me down and how I find it peaceful to just watch his slo-mo movements. He says he shares my experiences. "I'm surprised by how they move slowly but they move fluidly. It's interesting to see how they act in the water. They don't make noises—they just move. They just move around all day." He finds it meditative. "You can just sit there and watch them. They don't make a noise and you don't make a noise. You're with your thoughts. You wonder how they live their whole life in a situation like this," he points to the tub, "where they're in here isolated in this created habitat or out there in the world in their own habitat with predators, family, different animals, and logs in the water."

"She's in here in this tub with nothing but water," I make the obvious comparison slowly—I don't want to offend, "and you're in here in this prison."

"Yeah," says Hill. "How ironic. That's something I look at all the time. She didn't ask to come here," another laugh, "even though she didn't commit a crime. It's kind of a similar situation. I think about that. I'm incarcerated and I'm taking care of an animal that's isolated. It's something I really reflect on, I know that." He hopes that the treatment he and his incarcerated colleague perform for the turtles, along with their close observation of them, helps researchers diagnose what's causing the shell sores, how best to treat them and how to prevent the lesions from occurring. "Our ideas and suggestions are embraced," he's understandably proud to report. We look at the swimming turtle. "You have to have patience," Terrell Hill says as we watch the turtle, and I ask him again why such mundane activity is so captivating. "That might be the million-dollar question," is his easygoing answer.

LOVE MY LIFE is tattooed in flowing cursive along the inside forearm of Joseph Goff, the western pond turtles' other caretaker at Larch. Goff sports a hipster's few day's growth of beard and eyes that smile in concert with his frequent grins. He grew up around Lake Tahoe, surrounded by wildlife. "I have a passion for animals, and it is a pretty cool opportunity to be able to take care of the turtles." He, too, sees parallels when he considers his current lifestyle and theirs. "We inmates are in an enclosure. We're getting better and then we're being released." But people in general like turtles, he figures, because they're easy to take care of and they're friendly.

"Some don't seem too bright," he candidly critiques the Larch turtles when I ask him why he thinks turtles are idolized as totems of wisdom. "We had a couple that were just kind of dopey and focused on just eating." But there isn't much for them to do in their isolation tanks—there is nothing in the tanks except for basking pads and a floating shelter. "Yeah, it's kind of boring for them." But not for Goff. He and Hill note the turtles' behavior and write reports on their overall appearance. They check on things such as the condition of their feet and the color of their skin and shells and make sure they're BAR: bright, alert, and responsive. Zoo-keeping tasks. If they see something amiss, their concerns are reported to the Portland Zoo veterinarian in charge of the project. They study western pond turtle anatomy and read research papers to help them better understand their charges. They raise mealworms to feed to the turtles. Goff is pleased to be learning caretaking skills that can transfer to jobs on the outside. "It brings a new kind of appreciation for something other than myself. That's nice." He also finds himself meditating with the turtles. "You've got the quietness, away from everything." The turtle lab is separated from the bustle of the other prison buildings. When he's watching them, watching what they're doing, Joseph Goff says he's not necessarily thinking about the turtles but rather about life in general. "Just slow down a little," is his message. "I want to give back. I've thought about that over the past year. A lot of people need help, a lot of animals need help. This has definitely changed my focus. When I get out, whatever I do, it's all going to be toward helping others, helping other things." Quite a gift from the turtles to their caretaker. "Yeah," he agrees, "no matter how small they are, there are other things that need help."

Before leaving the prison I check in with the Larch correctional program manager, J. C. Miller, who oversees the western pond project. "The turtle program gives offenders something to believe in," he says. "It gives them something to own. It gives them responsibility." When Terrell Hill and Joseph Goff are released from Larch they'll leave with a certificate acknowledging what they learned and the work they accomplished. "And it helps the turtles. It's a win-win, rehabilitative for the offenders as well as the animals. It helps both ways because something needs you, and I think we all want to be needed." Miller had a turtle himself when he was growing up, but when he got a boa constrictor (burly and ex–air force, Miller looks like a guy who would be comfortable with a fat snake around his neck), the turtle became just a memory. That boa was his favorite reptile, with an endearing personality he was convinced no turtle—western pond or otherwise—could match.

• Fast Fred •

Fred and I commune this morning, at least with some reptilian rapport. I look in on him when I come downstairs to the living room and he's flopped in his bathing-drinking pool with his hindquarters hanging out up in the air and his front claws in the water along with his mouth. He looks up at me and when our eyes meet we both hold the moment. He blinks. He turns and climbs out of the water. But then he stares back at me again. I talk to him. "Hello, Fred." And, "Good morning, Fred." He blinks. But he looks at least curious. Can Fred be curious? "Hi, hi!" I say and he stretches his neck out toward me. Maybe his stare is just primal hunger combined with hope and expectation—nothing personal. I am a recent source of food, a potential source of more food.

One thing for sure: Fred is not always slow. A sudden move from me as I study him and his head darts in a flash back into his shell. I move to look directly at the front of him so that I am looking into both of his eyes at the same time—usually Fred just gives me his (best?) profile. He looks directly at me. No question he knows I'm there and he is relating to my presence—but how? Enjoying my nice, soothing voice? Worried I'm going to grab him and make soup? Wishing I would change his water? Just wondering why I keep looking at him? As I stare back, it sure feels to me there is some contact happening. We're connecting. That reptilian brain knows I'm here, and my reptilian brain—along with new and improved gray matter evolution blessed me with—is trying to relate to good ole Fred. But perhaps it's time for me to be more Fred-like when we're together. Perhaps it's time to take a break from trying to analyze his moves and his intentions and our relationship. Perhaps the journalist should stop trying to report on Fred as if he were a news story and just hang out with him for the sake of hanging out.

SIX

THE ILLICIT HUNTS

— Hunting on the Bayou —

Morning on an already hot Louisiana swamp. The still water ahead is disturbed only by the nose of an alligator crossing in front of the flat-bottomed aluminum boat ferrying me fast toward it. Or am I imagining that nose and it's just a knot on a rotting log? I'm on Bayou Teche with turtle and alligator hunter Russell Bourgeois. We're off turtling in Louisiana sugarcane and oil country—famous for its zydeco music, crawfish étouffée, unique French patois, and intimate relationships with the outdoor wilds. Especially turtles. Now retired, silver-haired and tanned from his days on the bayous, Bourgeois savors "the thrill of doing something not a lot of people can do." He used to catch lots of alligator snapping turtles in these waters—catch them, weigh them, measure them, determine their sex, tag them, and release them with hopes that he'd catch the same specimen on another outing as part of his ad hoc research into the status of the bayous. These days he's worried about the sharp decline he's witnessing in the overhunted alligator snapping turtle population. He spends some of his retirement hours searching for eggs and incubating them, and when they hatch, he escorts the babies to locales where his experience suggest they'll most likely survive. Not that he shuns turtle soup. "The common snapper and the softshell," he tells me with his easy drawl, "I've got no problem killing them for table fare." But he understands why some of his neighbors poach alligator snappers. A hundred pounds of turtle yields about thirty pounds of meat, and at $8 a pound, pulling just one hefty turtle out of the bayou is a good payday, especially in rural Louisiana.

The Evinrude 250 outboard whines loud, speeding us along at a steady twenty-five-miles-an-hour clip—fast enough to keep the deerflies from bothering us. We veer off into a tree-covered canal dug years past for a pipeline. Bourgeois cuts the motor. We've found the first of the jug lines he set in the water the day before. His Styrofoam buoys mark tough cord attached to a hefty steel hook that's baited with catfish. "They're an ambush animal," he explains, pointing to the cypress along the canal bank, roots disappearing into the water. "They like to sit undercover on the bottom." Cicada choruses sing back and forth from the tree-tops competing with the grunts and groans from Bourgeois as he tugs on the line. "We might have a turtle here." He doesn't like the feel of the line. It's snagged. "I hope it's not drowned."

The turtle breaks the surface, dead. Alligator snappers must come up from the bottom for air, and, once hooked, this poor guy was stuck underwater, tangled in the line. Bourgeois is crushed. He's not just lured to the water by the thrill of hunting turtles. He mourns any unnecessary loss of the exploited species, an animal that must live some dozen years before it begins to reproduce. He knows greedy poachers plague the swamps, dodging game wardens. "Like speeding on the highway, if they don't catch you, you get away scot-free." In Louisiana, the law is lax: One alligator snapping turtle a day per boat is legal to take, but since the meat is illegal to trade it's hard to imagine what anyone could do with 365 turtles a year besides sell or barter most of them. The few alligator snapping turtles Bourgeois keeps are in ponds on his property. He figures otherwise there's a good chance they'd be dinner elsewhere. He hefts the dead turtle and unhooks it. "I'll put him on the bank and a bear'll eat him." He looks at the lifeless link to the dinosaur age, estimates it's about ten years old. "Unfortunately that's the chance you take when you're doing this." That ends our legal hunt. Louisiana's one-a-day rule counts turtles dead or alive.

Another trip on the bayou. We putt along at a crawl toward a buoy, Bourgeois slapping deerflies off his head. (I've sprayed myself with poison and they leave me alone.) "This is one place you would not want to be in the dark unable to get out. The mosquitoes and everything else would eat you alive." This time we and the turtle are lucky. Or at least we are. Bourgeois yanks on the line and up comes a monster, thrashing the water with its powerful legs, its beak ready to chomp. He estimates thirty pounds of turtle and gets a net sack ready to bag it. "Grab his tail!" he orders. "He wants a piece of my behind!" Although it flails about, hanging from my clutch above the water, it can't reach Bourgeois while he maneuvers it into the netting. It is one heavy load, its coarse and damp-with-swamp-scum tail in my two-handed grip linking me in that moment to the dinosaurs. Or so I feel. The turtle's mouth is wide open. "He's mad," says Bourgeois, "and he obviously doesn't want to get caught. He's doing everything he can to protect himself and escape." The hook stays in the turtle—yanking it could damage its innards. "Its digestive juices are so strong that within a couple of months it can dissolve the hook." Dissolve the steel hook? "Oh, absolutely, yes, it will."

Bagged and onboard, we take a closer look at the "loggerhead"—Cajun jargon for alligator snapping turtles, which can be confusing since this is also the common name of *Caretta caretta*, the sea turtle that lives just a few miles distant in the Gulf of Mexico. Bourgeois points at a pink appendage just inside the mean open mouth. "See how he's moving it? If he was in water, that thing would be wiggling. Fish come by thinking it's a worm and try to get it. When they do, they become supper." What's the turtle thinking, here in the boat with us? "He's just thinking he'd like to be back in the water." Bourgeois muses that he'll probably return it to the bayou after watching it for a spell in a tank.

There's nothing Fred-like about the monster sharing the boat with us, nothing sweet and endearing. It's mean-looking, mean-acting, and ugly. When it yaws its mouth wide the message is clear: I want to hurt you and maybe eat you. Its pointed beak is like a raptor's—designed to cause damage. Its claws reach what looks like a couple of nasty inches out of its gnarled leathery feet. "Razor sharp," warns Bourgeois. The beady eyes and the weathered gray bumps and ridges on the carapace of this *Proganochelys* cousin add to its sinister prehistoric presence. "I've always had an affinity for these turtles because they've been exploited. I guarantee you 99 percent of the kids in south Louisiana would put that thing on the table." And shellac the shell for a trophy.

A few minutes later we find another and bigger turtle hooked. The first catch goes back into the bayou and Bourgeois wrestles number two onto the gunwale of the boat, upside down with its head hanging over the side. As long as its head is suspended in the air it can't leverage itself off its back. It stays immobile while Bourgeois nets it. More grunts and groans. "Man, these things will give you a workout." Sweat pours off him. The sun is rising higher, along with the temperature and the humidity. Hawks patrol above the trees, snow-white egrets glide along the water, moss hangs low from the cypress, lilies and dragonflies decorate the water. My guide heads the boat into brush along the bayou bank and warns me to watch for wasp nests. "Those things'll light you up and make you holler for your mama!"

Another jug line. Another turtle? Not this time. It's an alligator. Seven feet or so, Bourgeois estimates as he cuts the 'gator loose. "I have no fear of them," he says about the snapping turtles and the alligators, "but I have a huge respect for them because I know what they can do." The beak of the turtle in the boat resembles the hawk above us. "Whatever they bite on," he says about alligator snapping turtles, "they're gonna keep.

They're not super agile. They're not very fast. They're just extremely powerful."

We check another dozen or so jug lines. No more turtles and no bait left on most hooks. "Thieves," Russell Bourgeois mutters as we head back to shore with our catch, and he's only half joking about the anonymous varmints that took his bait.

— *Turtle Poachers Busted* —

Tourists drink in the frivolity of the New Orleans French Quarter not far from Louisiana's darker side. West out of the Crescent City, I-10 cuts a straight line through swamp country, through Cajun country. Just off the interstate life is often hardscrabble, and the bounty of the bayou is a viable supplement to the grocery store. For those tempted to break laws, the alligator-filled swamps are virtual ATMs. "Poaching is just a way of supplementing their income," says FWS Special Agent Jim Stinebaugh. "Poaching supports the lifestyle they want to lead." Among the easily available booty are alligator snapping turtles, still plentiful on the Texas side of the state line, where they're listed as threatened and consequently are illegal to mess with.

Special Agent Stinebaugh looks the part of a Western lawman, from his steely eyes and chiseled jawline to his two-toned cowboy boots tucked under straight-legged jeans. Tommy Duncan (of Bob Wills and the Texas Playboys fame) is singing through the speakers in his office when we sit down to talk soon after Stinebaugh and his posse busted a poaching ring accused of yanking at least three score alligator snapping turtles out of Texas waters and smuggling them into Louisiana. Butchered and sold would have been their likely destiny had the feds not intervened.

The tale starts as so many crime-stopper stories do: with a simple traffic stop. It was 2009 when city police pulled over a couple of good ole boys trailering a boat and speeding through

Jasper, Texas. When the cops spotted a dole of alligator snapping turtles stuffed in the boat, they called the county game warden. He sauntered over and asked what was up. "We were over here fishing for catfish," the warden recounted their response, "and we happened to see all these turtles running 'round. So we picked 'em up and decided to take 'em home. In Louisiana," they patiently explained, "we eat turtle and we're gonna split 'em up and eat 'em." The warden wrote two tickets for the illegal take of the threatened species and released the turtles into a Jasper County creek.

A couple of months later Agent Stinebaugh lunched with the warden and heard about the haul. "I've always liked turtles," Stinebaugh tells me when he and I meet years after that meal, "and I think a lot of people like turtles. They're just fascinating creatures. They're dinosaurs, especially alligator snapping turtles, true dinosaurs." After the lunch with the warden he spread the word to county wardens all along the Texas-Louisiana border asking them to get in touch if they caught any turtle poachers. In 2013, one called and said, "Hey, I just caught this guy with a bunch of turtles and he's from Louisiana."

Taking a protected animal from one state into another, even if it is not illegal to hunt in the destination jurisdiction, violates the federal Lacey Act. That means hauling those Texas turtles across the state line is a federal offense. Stinebaugh headed out of Houston to meet the smuggler, one Joseph Guidry, "an unemployed welder and handyman, mainly just a poacher." Stinebaugh knocked on Guidry's door and his common-law wife came out wielding a butcher knife asking why he wanted to talk to Joe. He told her "about some trouble in Texas." She told him to come back later, and when he returned that evening he met Joseph and his father waiting for him in Guidry's little trailer, the two sitting in a haze of cigarette smoke. "I told Joe he was in a lot of trouble, that the reason Texas let him go was to turn him over to me. 'You've got two options,' I said, 'Work with me or you're gonna go to jail.'"

Sitting on the table in the trailer was a stack of hundred-dollar bills Stinebaugh had pulled out of his pocketbook, what the feds call "purchasing-information money." Guidry looked first at the cash and then at his father, who said, "Just tell him what you know, Junior." Junior spilled.

"I'm catching these things and sellin' 'em, but a lot of other people are doing it too. And they're doing it a lot more than I am." Stinebaugh's response to Guidry was simple. "Tell me who these others are," said the agent. Names were not all the poacher provided. Stinebaugh wanted him to wear a wire and chat up his confederates while the feds recorded them incriminating themselves. Guidry balked at first. He told Stinebaugh that the other poachers were dangerous, that they would come after him and hurt him if they thought he was doing something to hurt them. "This is your one chance," Stinebaugh replied. "We can prosecute you and make an example out of you. But you're telling me there are bigger fish out there. I'd rather get them." Guidry agreed to play informant. "I don't know if it was the money that sold him or the idea of getting out of trouble. Maybe a combination of both."

Joseph Guidry proved ideal at extracting information from his unsuspecting friends and neighbors. They recounted in boastful detail their poaching exploits, unaware while they spun their tales that the feds were recording every word. "I'm gonna get a beer real quick," says a voice on the recording, and ice in a chest rattles as Guidry encourages one of the smugglers to brag, musing that he heard "a month, a month and a half ago, y'all caught a bunch, heard you guys caught like twenty or somethin'. Somebody was tellin' me that."

"Not me, buddy," says the suspect. "I wish. I caught eight."

"That's the most you caught?" Guidry pushes.

"Yeah."

A car passes, crickets—or maybe it's cicadas—chirp. This is an unhurried conversation. But the other guy takes the bait and drawls about his exploits. "It was Mama's little car," he says about transporting alligator snapping turtles, "ended up burnin' the head gasket on it and I had to get her another one." Another head gasket, apparently, not another car. "I fixed it and shit." But soon, listening to the slow-talking suspect pays off. "Another time I bought a car and I rigged a trailer hitch on it. I had so much frickin' loggerheads in the trunk and everywheres, and the boat was loaded down. I tried stoppin' at a stop sign, it pushed me clean across a main highway." The two guffaw. "The smoke was flyin', couldn't stop that load. The li'l car was too light."

"Cool," says Guidry, and he waits for more detail. The two talk fishhooks and the power of a loggerhead's bite. "My damn tomatoes are all rotten," says the suspect as they walk around his yard. "See that one there's gettin' a brown spot." A car rolls past, tires humming fast on blacktop. Birds sing. The informant's patience pays off as the suspect starts talking about haunts in Texas thick with turtles, about the cost of fines when local game wardens catch turtle poachers.

Guidry seizes the moment. "When y'all fishin' turtles back in Toledo Bend, how many would you reckon over hundred pounds would you mighta caught outta there? Just a couple?" The lake straddles the Texas-Louisiana border.

"Probably five hundred," comes the quick response.

"No shit! That many? Over five hundred of 'em?"

"There ain't nobody gonna ever catch as many turtles as I caught. You could wipe out all the rest that they got in Louisiana and Texas and never have as much as I caught." This is a poacher who knows how to catch turtles, knows what he is doing is illegal, and knows the harm he is causing to the loggerhead population.

"That ain't hundreds of pounds, that's tons of turtle," muses Guidry.

"Oh, yeah," comes the confirmation.

Still slow-talking, Guidry guides the suspect through further stories, successfully encouraging him to name names of other poachers and the bodies of water they pillaged.

Agent Stinebaugh's investigation took a leap forward with the information Guidry captured. He determined that supply and demand lured the Cajun turtle fishers across the state line to Texas. "Louisiana is pretty well fished out," the investigator tells me when we talk in Houston. "You can still catch 'em over there, but if you want to catch a big turtle, if you want to come home with a boatload, you go to Texas." Cultural differences help drive the trafficking. "We don't eat 'em here," notes Texas-raised Stinebaugh, "so nobody's going to be gunning for 'em here."

Police work can be a slow trudge, and Stinebaugh spent a couple of years—when his other duties allowed him the time—pursuing names and places identified by Guidry. "I believed firmly that there was significant damage being done to the re-source, to this animal that's very near to being on the Endan-gered Species List." At this point in the ongoing investigation, the turtle case turned violent. Joseph Guidry's common-law wife—the one who had greeted Stinebaugh at her trailer door with a knife—called Stinebaugh in the middle of the night, panicked. "He's dead, he's dead," she cried to the cop. "He's killed himself. He just put a shotgun to his head and pulled the trigger."

Why did he commit suicide? Guidry suffered from physical ailments, according to Stinebaugh, and from drug and alcohol abuse, "But he did express to me on several occasions that he was scared the people he was getting information from were going to come and hurt him." Regardless of his motive, the shot-gun blast that killed him destroyed the courtroom value of all the evidence he gathered. "Everything's gone because he's not there to bring it in. I couldn't because I was not privy to it."

Among the people of interest listed in Stinebaugh's notebook was perhaps a distant relative of Joseph Guidry, one Viola Guidry, a woman he called Aunt Viola. Stinebaugh surveilled her sprawl-ing property—a house and big ponds, all surrounded by stout cyclone fencing—and what he saw convinced him she was conducting illicit turtle business. Through a series of lucky breaks and with zealous gumshoeing, he made contact with Viola's boyfriend's brother (many of the people in the small towns of Calcasieu Parish are closely interconnected by friends and relations). The ponds at Viola's place were full of wild-caught alligator snapping turtles, the brother confirmed. He agreed to sell one to the feds' informant and deliver it on the Texas side of the border.

Initial negotiations were captured via a police phone tap.

Informant: If I could get ahold of something like 120 pounds, that would be just about perfect.
Poacher: We have a 121-pounder and a 145-pounder.
Informant: Let me get the 121-pounder.
Poacher: Uh-huh, yeah. Would you like us to clean it?
Informant: Clean it and get the shell?
Poacher: Yeah, we clean it. We charge $25. We'll give you the shell if you want.
Informant: How much is the shell?
Poacher: The shell's free if you buy the turtle. The shell itself sells for fifty bucks once we clean 'em and shellac 'em and everything.
Informant: Let's go ahead and do that. So what am I lookin' at, total cost?
Poacher: Two-fifty a pound for the total weight of the turtle.

That's a $327.50 sale. It and a few more recorded turtle negotiations were enough to convince a judge to issue a search warrant of Viola's spread.

"We ran hoop nets for three nights," Stinebaugh says about the unorthodox expedition at Viola's ponds, "and we caught thirty turtles." But when alligator snapping turtles get spooked they tend to hide in bottom mud. "It's possible some of 'em just went deep and we never saw 'em again." Day and night Viola watched the police action. "She was very calm. She confessed." She confessed to selling poached turtles and she confirmed that their provenance was Texas. One of her clients was Jimmy Mistretta, owner of the Loggerheads Bar and Grill on the bank of the Calcasieu River in Lake Charles. Stinebaugh visited him as Mistretta was preparing to kill two gigantic alligator snapping turtles that he had bought from Viola, a 165-pounder and the other weighing in at 171 pounds. "He was gonna ax 'em, mount 'em, and put them up on display in his saloon," Stinebaugh told me. When I stopped to look in on the scene at the bar, two other alligator snapping turtle shells, relics of much smaller specimens and glistening from shellac, were hanging on a wall near the toilet doors. Owner Mistretta played food critic for investigators about the taste of his joint's namesake. "The meat's the best meat you've ever had in your life," he rhapsodized to Stinebaugh. "Never tasted anything better."

As Viola's confession continued, she explained how poachers sell protected turtles at local de facto black markets by labeling them as common snapping turtles, since savvy buyers recognize them as alligator snapping turtles. "Of course she's depressed—we're serving a warrant on her house," Stinebaugh observed at the time. There were guns in her place, guns secured during the search. But once the operation was completed, the agents packed up their gear and the evidence they seized, and they returned the weapons.

A few days later Stinebaugh was sitting in his front yard relaxing with a cigar, quite pleased with the latest results of his turtle operation, when he received another unexpected report that temporarily derailed the case. "It was one of those guns she killed herself with. Or that she was killed with." Stinebaugh, for one, is not convinced Viola died by her own hand. But her death was ruled suicide by the parish officials, authorities who reported the deed occurred in the doorway of her church, that she was killed by a shotgun blast. No matter how she died, with her went the value of her confession as testimony against the other defendants Stinebaugh was corralling. Two dead. Over turtles? "We started again to build a case. Not with a blank slate, but with something close to a blank slate."

Build a case they did, and in the spring of 2017, they charged four of the gang with felonies: conspiracy, violating the Lacey Act when they crossed state lines with sixty-six poached turtles, destroying evidence, and lying to a federal officer.[1] Three pleaded guilty to felonies and one to a misdemeanor; two of the felons were sent to federal prison, one for twenty-one months and the other for sixteen months; the third received probation. The poachers themselves valued their take at as much as $150,000 on the retail outlaw market.[2] Stinebaugh hopes the drama of the case will help list alligator snapping turtles as endangered and that it will influence Louisiana lawmakers to toughen their state's take rules of one alligator snapping turtle a day per boat. He expects the rustlers he caught are not the only Louisiana lawbreakers preying on Texas turtles. "Louisiana law is terrible because it allows them a perfect excuse to be in possession of turtles. They can just say they took one a day. You could have fifty at your house because there is no possession limit."

But the conservationist lawman tries to understand the lawbreakers. "It's cultural," he sighs. "These guys grew up eating turtle. It's part of their identity. Alligator snapping turtles aren't easy to catch or come by." The poachers he describes can seem like folk heroes. "If you were to meet them they would be quite charming. They're supplying a product that's in demand. These guys don't think much of it. It's a turtle." As we talk he imagines the point of view of the poachers' customers. "'I've been eating turtle all my life,'" he says feigning their voice. "'My granddaddy ate turtle. My dad ate turtle. Now I eat turtle. I could go catch it myself. They're just catching it for me.'"

Not that he excuses the crimes. "The poachers are bad guys, for sure. They need to be stopped. They're willing to go out and do significant damage to make a little bit of money to supply their need for drugs and to supplement their lifestyle." He generalizes about the types of men who become poachers with a crushing and judgmental litany. "No work. Move in with the girl that's getting some public assistance. Live in the family trailer house where you don't have to pay rent. A little bit of theft. A little bit of drug sales. Criminals."

"You're painting a nasty lowlife picture," I say.

"Very much so," he agrees. "Very much so."

One more question is on my mind before I head off into the Texas sunset remembering the postcard line, "The sun is riz, the sun is set, and we ain't out of Texas yet!" I want to know if, after chasing turtle stories for years across Cajun country, Jim Stinebaugh has ever eaten turtle.

"Never, but I've got one in my freezer," is his measured response. It's the first 121 pounds of alligator snapping turtle bought as part of the undercover investigation that shut down the Calcasieu Parish bunch. "I'm considering having one of our agents cook it up when we get rid of it. We're gonna have to throw it away. There's nothing else to do with it and I'm curious."

"We should probably have that meal together," I suggest, telling him about my own tussle with whether eating turtle is required of me as an element of my research.

"I want to see why people are interested in poaching them," he says. "I certainly would not kill one for the purpose of eating it. But we have one that is already dead, and I'd like to know what all the fuss is about. Why are people willing to break the law to get these things?"

Or as Tommy Duncan sings with Bob Wills and those Texas Playboys accompanying him, "I don't worry cuz it makes no difference now."

— *Modern Moonshiners* —

In the relatively recent past, turtle poachers and smugglers worldwide sought wealthy collectors in the United States as customers for the rare, protected animals they sold on the black market. Rich Americans (and Europeans and Japanese) were their primary customers. Customs inspectors and wildlife police watched for suspicious incoming parcels and for arriving passengers nervously exhibiting strange clothing bulges or suspect suitcase compartments.

But in the last several years a radical change has been occurring—one centered on globalization and the growth of the middle class in Asia. Turtles are the victims of this phenomenon, becoming leading characters in an international show starring a highly unlikely cast of intercultural coconspirators.

Outdoorsmen (and they *are* mostly men) in the hollers and bayous of the southeastern United States are fueling the little known and skyrocketing export-smuggling trade in turtles native to North America. These trappers and traders use their intimate knowledge of the rural Southeast to catch the turtles (box turtles, mud turtles, and painted turtles are among the turtles prized abroad) and then sell them to smugglers who use their language and culture skills on both sides of the Pacific to illegally export the turtles from the United States to anxious Asian markets hungry for the chelonians.

Business is booming for the U.S.-based smugglers, to the point that they cannot keep up with the demand. Law enforcement sources report that 50 percent of the FWS turtle inspection work now is directed at export smuggling. "All reptile cases back in the early 2000s were incoming." I'm told by my FWS sources, "and lots of them turtles." That dynamic changed by 2009. "By late 2015, we were regularly working on turtle smuggling cases involving large quantities of turtles leaving the U.S. for China. The import smuggling still happens for the pet trade, but the vast majority of turtle smuggling shifted to outbound smuggling." Native American turtle populations are in immediate jeopardy.

The culture clash of this new development finds rural Southerners in unlikely cahoots with urban Chinese-Americans and

Chinese nationals. The reversal of the historical illicit wildlife trafficking routes is counterintuitive. It's as if Oregon marijuana growers were selling cannabis in Mexico.

Stealing turtles from Southern bayous, ponds, and streams to make a quick buck selling them to faraway foreigners creates an ecological disaster (and one replicated by impoverished natives throughout the world). Bales of turtles are captured and shipped illegally to eager buyers annually. Stacks of tax-free turtle dollars are changing hands each year, and that figure keeps growing.

• Stress-Free Fred •

Today is a bright sunny autumn Pacific Northwest day outside and a bright day for Fred under his basking light. It's been a busy caffeinated week at the university, not enough sleep and too much running around trying to cram two days' work into each day. I look in on Fred and soak up his pace as a stress antidote. He's stretching his neck out, pushing his head as close as he can to the lamp's heat. Slowly. He looks at me with his big (for his head size) brown eyes. Brown eyes mean he probably is Fred and not Fredericka. I look back at him with a "Hey, Fred" greeting. He doesn't mind how close I get as long as I make no sudden movements. If I do, his head snaps back for a second while he makes sure I'm no threat. I had thought it might soon bore me to watch him, try to commune with him. But I'm finding time with him restful, peaceful. Watching his slow, deliberate movements calms me and removes me from daily mundane concerns as I keep wondering what he's doing and why. And if he's thinking, about what?

Breakfast time and it's a banana and lettuce for Fred. He likes lettuce okay and carrots more. Shredded carrots. But of the three, bananas are the runaway favorite. This morning he wastes no time clambering over his newspaper (wet in places—I must change it today) to his food saucer. He steps over the lettuce—ignoring it—and attacks the banana with the same neck-stretching snap he uses on his worms. That technique results in mashed banana on his face, a mess he wipes off with the backside of his right foot and then his left, like a cowboy at the chuck wagon. Enough, he figures after several mouthfuls, and abruptly he turns from the saucer and marches through the archway into his dark room to digest.

— *Urban Turtle Communing* —

While I was in Vancouver's Chinatown researching the butchering-a-live-turtle case I stopped by the Dr. Sun Yat-Sen Park. It promised to be an appropriate oasis to contemplate the gruesome marketing of turtle meat. Its pond is home to turtles that need fear neither smugglers nor a butcher's knife. As I entered the garden on a late-summer midday, I scanned the lily-covered pond and saw two turtles sunning themselves on a rock just a few feet from my vantage point. The smaller one placed a paw on his buddy's shell-shoulder. Both their heads were outstretched, soaking in the sun's rays.

"They're trying to push each other off the rock," a passing tourist reported to his companion.

I don't think so. To me it looked like a friendly embrace. A fat goldfish or koi carp rollicked past the pair. The turtles looked up at me, their eyes bright. The one with his foot on the other winked. Did that really happen? Without question, we were inspecting each another. As I watched them, the city noises—an elevated train, car horns, hammering at a construction site, a passing airliner—all started to fade for me in favor of the garden's quiet. The winker climbed up on the back of the more sedate turtle. Slowly. But not pushing. These two, I am convinced, were just hanging out (and still staring at me). I needed to leave them. I had an appointment across the border later that afternoon, and the delay at the crossing can be nasty. But funny: I didn't want to leave. I wanted to stay with them longer—us contemplating each other certainly relaxed and refreshed me.

I silently said good-bye. Maybe it was also a telepathic good-bye. I hope so.

• Fascinating Fred •

Fred just yawned. Right now he is one relaxed turtle, sprawled out on his belly, his legs splayed, his neck extended so far out it's looking almost snakelike, with his chin resting—like a publicity photo pose—on his heating pad.

SEVEN

THE WILY SMUGGLERS

— Turtles Stuffed Down His Pants —

Kai Xu headed into the Detroit-Windsor Tunnel undoubtedly nervous, checking his rearview mirror. The twenty-six-year-old college student had made the crossing plenty of times before, but on this trip he was not alone. Riding with him were fifty-one turtles from Louisiana destined for Canada and probably Asia. Xu was a turtle-smuggling middleman, buying from American breeders and selling to China, where customers waited, happy to pay Xu's high prices for a commodity illegal to ship across the Pacific without a federal permit. Gaining a permit would have been quite the challenge for "Turtle Man"—Xu's nickname assigned him by FedEx and UPS workers—since the fifty-one turtles were poached.

When I talk turtles, as I'm wont to do since hanging out with Fred and researching his kin, the sad tale of Kai Xu never fails to grab attention—even from the most cynical and blasé. The sad tale of turtle smuggler Kai Xu and the turtles he stuffed down his pants.

"He did what?" I'm asked. "He stuffed them where?" Next come the jokes about what the turtles might do to Xu's anatomy. Headline writers loved the story. After Xu was arrested, and denied release from jail on bond, *Detroit News* reporter Robert

Snell blared, "An alleged turtle smuggler faces long stretch in a shell."

U.S. Fish and Wildlife Service special agents had been watching Xu, a naturalized Canadian citizen born in China, for several months before they received a tip early in August 2014 from a Detroit UPS worker (who wanted his identity shrouded and was known during court proceedings only as Dave). Dave called authorities because a shipment aroused his suspicion. It was a carton addressed to Xu for pickup at the Hoover Street UPS depot, and it was marked LIVE FISH KEEP COOL— unusual instructions for a routine parcel. Agents staked out the UPS site. They figured Xu was en route. Border agents in Windsor had alerted them when Xu entered the tunnel— in a tan Ford Escape with Ontario plates—heading for the States.

The Escape, with Xu at the wheel, showed up that same afternoon at the Hoover Street UPS. Agents watched Xu retrieve his box and return to the back door of his Escape. From their vantage point they could see Xu open the box and repack its cargo into plastic baggies. They noted that he fetched a plastic grocery store sack from inside the car, took a new roll of clear packing tape from it, and "removed the plastic 'self cutter' on the roll of tape and discarded it and the wrapper on the ground under the Escape." Add littering to his alleged crimes. Xu didn't need the cutter; he was carrying scissors.

Next agents watched as Xu transferred the tape and the baggies—baggies that now appeared to be filled with something—into the grocery bag. He left the back of the Escape and strode toward a row of parked UPS trucks. "Xu continuously looked around the parking lot as he walked," the agents reported. At that point he disappeared from their view, spending about ten minutes between two long UPS trailers. When he reappeared, "again glancing around the parking lot," there are new and noticeable "irregularly shaped bulges under Xu's sweatpants on both his legs."

Agents tailed the Escape with Xu again at the wheel. He parked briefly near a Hoover Market debris box (where agents find the discarded UPS carton addressed to Xu) before heading back into the tunnel and out of sight. Canadian officials were notified that he was heading toward them. On the Windsor side they waved Xu into a secondary inspection lane and ordered him out of the Escape. A quick search revealed the fifty-one turtles taped to his body—forty-one to his legs and another ten taped around his groin. Eastern box turtles, red-eared sliders, and diamondback terrapins.

The Canadians confiscated the turtles and sent them for foster care to the Detroit Zoo. Border patrol officers on both sides continued to monitor Xu, who crossed back and forth between Michigan and Ontario several more times before he was arrested for attempting to ship two suitcases full of turtles to Shanghai.

"Mr. Xu is a fulltime reptile smuggler," argued Assistant U.S. Attorney Sara Woodward, convincing the court to deny his release on bond while his case was pending. At the bond hearing, FWS Special Agent Mona Iannelli testified that a "confidential informant" triggered U.S. agents' interest in Xu about six months before they arrested him, and she provided detailed examples of his smuggling operation—trafficking the government estimated was worth more than $1 million.[1]

As detailed in the federal indictment, Agent Iannelli had watched Xu conduct a series of turtle shipment escapades. In Clinton Township, Michigan, he picked up two packages from FedEx, after which Iannelli engaged in a "rolling surveillance," following him to a UPS office, where Xu picked up a third parcel. She watched as he opened them in his car. "We saw him discard some of the packaging material in the UPS parking lot"—again, official disgust with littering. "He just threw it on the ground." Next she watched him get out of his car carrying a pair of boots, which he took into the UPS office. He was still carrying the boots when he left UPS. He put them back in his car and drove off.

Kai Xu was not the most sophisticated smuggler sending turtles to Asia.

The boots were filled with something, a clue the UPS worker told Agent Iannelli that he determined because of their weight. But his office did not have a box in stock big enough for them. So Xu drove to another UPS drop-off station. The clerk there also rejected the shipment, telling Iannelli he figured the boots were stuffed with drugs. Why was he suspicious? Xu told him that the boots were worth $100, that he wanted them in China the next day, and that he was okay with the $400 overnight shipping charge—a fee he said he would pay in cash. The UPS clerk refused to ship the boots and ordered Xu to get out of his shop.

More tales of turtle smuggling connected with Xu were told at the bond hearing—reports of rare Asian turtles spirited into North America and a variety of American species sent to Canada and Asia. One package intercepted by customs agents in Hong Kong contained snow boots filled with three hundred baby diamondbacks, among other species. Authorities assumed that some of Xu's packages had made it to their intended destinations, because he had sent shipments from the United States to China and there was no record of the Chinese authorities seizing them.

The most recent smuggling example entered into the court record occurred a couple of days before the bond hearing. Packages were waiting for Xu at a FedEx office in Novi, Michigan. Agents watched him collect the parcels and drive to a Clarion Hotel near the Detroit airport. He loaded the boxes onto a luggage cart and checked into two rooms. They watched him exit the hotel and followed him as he drove back to the tunnel. Canadian counterparts picked up the trail on their side of the tunnel and reported back to the Americans that Xu stopped at several ATMs to collect cash. He then picked up a confederate named Lihua Lin, and the two drove back across the border to the Clarion. The next morning they drove to the airport, where Xu dropped Lin off at the international terminal with two suitcases. Lin checked in for a flight to China and left the suitcases as checked luggage with the airline. Authorities confiscated them and, as Agent Iannelli testified, "We found the suitcases to contain two pairs of rubber snow boots consistent with the other boots we had seen."

Kai Xu was pulled over and arrested after he drove out of the airport; he had $10,000 in cash in his pocket at the time. Lihua Lin was nabbed before his flight took off for Shanghai. There were 970 turtles in his baggage—some packed in Xu's trademark rubber snow boots, others in boxes of Kellogg's and Post cornflakes.[2] He told agents he met Xu when he answered Xu's ad offering $4,000 for four days of work "selling cosmetics." When selling cosmetics morphed into smuggling live animals, Lin said Xu told him not to worry about the turtles in the suitcases because the authorities only look for drugs and guns.

Bond denied, ruled U.S. District Judge Donald Scheer. In court that day "Turtle Man" lost his swagger. "Boyish looking," *Detroit News* reporter Snell called him. "Boyish looking with black bangs hanging over the tips of his glasses." Restrained in handcuffs and ankle chains, wearing a red jailhouse jumpsuit, he sobbed during the proceedings. Months later, after pleading guilty to six counts of turtle smuggling, Xu faced a stiff penalty. Judge John Corbett sentenced him to just shy of five years in prison. Kai Xu said he was sorry for his crimes and thanked Special Agent Mona Iannelli and her colleagues "for stopping the darkness of my greed and ignorance." His lawyer, Matthew Borgula, was shocked at the sentence, telling me the crime would have been a misdemeanor in Canada and that Xu "thinks he would have had a fairer hearing in China."

The wood turtles Xu's procurers snagged are each worth $500 to the middleman. Once they get to China and other lustful marketplaces, they are worth three times that. The smuggler is a study in ethics versus profits. Is someone like Kai Xu immoral, amoral, or conflicted? And observing their modus operandi is intriguing: the ingenuity of some smugglers, the stupidity of others, and the absurd lengths some go to, to obscure the animals in otherwise legitimate shipments. One smuggler, for example, replaced the meat patty of a takeout hamburger with a turtle and rewrapped the package in its original KFC paper. Inspectors at the Guangzhou airport saw what they called "odd protrusions" sticking out of the hamburger bun, protrusions that looked to them like turtle limbs. "There's no turtle in there—just a hamburger," insisted the passenger, who tried to convince security not to search his stuff. "There's nothing special to see inside." Of course they looked despite his protestations, and they found—surprise—a turtle. The fellow did not want to travel without his beloved "pet."[3] Turtles have been found by customs agents in boxes labeled "books" and in shipping containers identified as fruit. Some live through their long journey; many die.

— *Turtle Smuggler Talks* —

I cross the Canadian border and drive south into rural Washington. I've secured an appointment with Nate Swanson, the pet store proprietor convicted of smuggling turtles to Asia. It took months to arrange the meeting. Swanson kept mum to reporters from the time the U.S. Fish and Wildlife Service agents raided his house and backyard turtle breeding facilities through to his plea bargain, sentencing, prison time, and release. "Everyone contacted me. I was in *USA Today*, I was in my hometown newspaper. I didn't read one of them. I didn't clip one of them. I didn't keep one of them. I didn't want to read it. I fucked up." Our final negotiations were via text messages, and after we talked in person I sent him a note asking why he chose to break his silence and talk with me. "To be perfectly honest," he wrote back, "I'm not sure why I agreed to it," and then he quoted an acquaintance he consulted while he was considering my request to meet: "What if he writes the story without you?"

The story includes a puzzled FedEx agent who called authorities when she encountered an animated carton. "I applaud the delivery service worker who first reported this scheme to law enforcement when she noticed a moving box and discovered a snake inside," said U.S. Attorney Jenny Durkan after Swanson was sentenced in 2014. "But for that discovery, these traffickers might never have been caught."[4] The fact that turtles might announce themselves in transit seems to slip past the conspiring minds of more than one smuggler. A Russian traveler with the novelistic name Innocent Kapustin was stopped at the Prague airport with forty-seven tiny endangered black-breasted leaf turtles (*Geoemyda spengleri*) in his grip. "The turtles were transported in very miserable conditions without air," a Czech customs official told reporters. Two were found dead, and the rest stacked "between turtle feces and rotting leaves." Officers stopped not-so-innocent Kapustin when X-rays lit up his bag and they "spotted the contents moving."[5]

For our meeting, I chose an anonymous freeway-side IHOP at an exit convenient for Nate Swanson and waited in a secluded booth with line-of-sight to the front door. Old pop hits were echoing through the near-empty pancake house. The Teddy Bears sang "To Know Him Is to Love Him," Eddie Cochran's version of "Summertime Blues" rang out, and I kept myself from chair-dancing to Billy Bland's "Let the Little Girl Dance." I listened for a tune appropriate to the moment. Elvis provided: "Return to sender," he sang, "address unknown, no such number, no such zone." And Nate Swanson walked in, his round face giving him a younger look than his thirty-eight years, his close-cropped beard and hair a good match for his schoolboy eyeglasses and untucked short-sleeved sports shirt. He sat down at the table, took out a pen and a little spiral notebook, and ordered only a glass of water.

"They were looking for somebody to crucify," he says about the case against him, "and they did. They couldn't have crucified a Chinese person." He figures the government wanted to showcase an American smuggler, since the Chinese nationals caught illegally exporting Swanson's turtles to China received six-month sentences while he got a year and a day. "I think you have to send a shot across the bow on your own home soil that this has to be stopped," is how Swanson analyzes the difference, "and you can't do that with a foreign exporter, with a foreign smuggler." Locking up a homeboy, he says, sends a stronger message.

Nate Swanson tells me stories from his boyhood, a time that sounds idyllic—life in the countryside mucking about on the family's five acres, growing up with farm animals and an orchard. A nearby pond proved particularly alluring for what started as a hobby. When it dried out in the summer, Swanson collected tadpoles and nourished them in tanks until the fall rains brought water back to the pond. "They would have died," he says about his rescues. Years later, in 2004, he tells me, a nearby worker unearthed a hibernating box turtle and gave it to Swanson. "He said, 'I don't know what to do with this turtle. Do you want it?' And I said, 'I don't know what to do with turtles. I'm a frog guy.'" But he took her home, convinced her to eat some raw chicken, and she thrived. "We still have her—her name is Sally."

Messing about in the pond was natural training for owning and operating a pet store. Swanson bred turtles, raised them, and then sold them. The indictment the feds presented to the grand jury calls Seattle Reptiles and Swanee's Exotics ideal fronts for Swanson's smuggling operation. Swanson tells me he opened the stores because he could make a business with animals he loves. He calls his operation an example of best practices, with the animals well treated, well fed, and well housed. Along with Swanson, five Chinese citizens from Hong Kong were listed as coconspirators, including one Leung Yan Fai, a reptile trafficker who also goes by the better-than-Hollywood nickname "Snaky." At about the same time Swanson opened his retail outlet, the indictment stated, he conspired with his Chinese partners to import and export turtles; the practice lasted for several years and violated both international treaties and U.S. law. They wrote fake labels for shipping containers and "falsely identified package contents on disclosure forms in an effort to deceive postal officials, wildlife inspectors and customs authorities,"[6] all without required import and export licenses. Cartons full of turtles were marked "boots and belts" and "T-shirts and used wristwatch" and "toys."

The indictment legalese describes dastardly international deeds, crimes that sound much more organized than the way Swanson defines those arrangements when we talk in the IHOP. "They found me by a classified ad," he says, about the Chinese nationals. Swanson, as do many in the animal trade, used the internet to solicit business far from the Washington base of his retail stores and its walk-in customers. On his first internet encounter, "I was selling two turtles, and they asked me if I could get ten. I said, 'Sure,' and they didn't complain about the price." He says those first sales were legal. He was sending box turtles from Washington to Nevada where two of his codefendants were enrolled in a Las Vegas community college. Schooling was their front, according to the government, for the smuggling operation.

But Swanson insists to me, as he nurses his IHOP ice water, that for years he knew nothing about the transshipment of the turtles from Las Vegas to Hong Kong. He says he did not know that the turtles he was selling to his Las Vegas clientele for $100 each were marked up to as much as $1,200 a turtle by the time they arrived at their final destinations. "And then, in the fifth year they told me that the shipping box needed to be quieter. I said, 'Why? What's the concern?'" The Vegas contingent told Swanson, he says, that there were problems after they sent it on to a city in China. "I say, 'Whoa!' and tell them why they can't send them to wherever this city was." He cites as reasons their lack of export permits, the excessive quantity that they're shipping, and the fact that they are not U.S. citizens. "Then they went dark for six months. Didn't contact me." But after those quiet six months another message showed up in Swanson's in-box. "We need you to ship us ten turtles," it said. "I don't know if it was a setup. Whatever." He dismisses the motive of that last communication with his partners. "It doesn't matter."

What matters is his response to their inquiry. "I did it. It was a moment of weakness. We needed to pay our mortgage." I ask him if he made that last sale knowing the turtles would be shipped on from Nevada to Asia. The answer is yes, and he did it knowing it violated laws. Next he swapped some of his turtles for turtles from Asia—black-breasted leaf turtles, an Arakan for-est turtle (*Heosemys depressa*) and a Bourret's box turtle (*Cuora bourreti*)—another transaction he knew was illegal. Meanwhile investigators were tracking email messages. "Snaky" wrote from Hong Kong confirming receipt of eastern box turtles, telling Swanson "the females are very nice, but the yellow male lose [*sic*] one leg." Investigators were tracking money transfers—$350 via PayPal one day from Hong Kong to Swanson, six weeks later an-other $630—never huge sums. The entire conspiracy netted him, according to the government, less than $200,000.

Why did he succumb to temptation, to violating laws he knew and understood? "I don't know, I don't know," is his repeated answer to my query. "Every time I did any transaction or any shipping I didn't feel good about it," Swanson says, looking away from me and off at the chain restaurant sameness of the IHOP décor. But perhaps at least a piece of the answer comes from Swanson's line: "Everybody wants what they can't have."

That's what U.S. Attorney James Oesterle figures motivated Swanson to break the law—that and the dollars. "I think a lot of it," Oesterle says about illegal animal trafficking, "is just the notion that 'I can get something that nobody else has.' It's that collector mentality. We've prosecuted trophy hunters who love the animal, but only if it's on their wall. It's pretty selfish think-ing that the only reason you like them is so you can look at them."

In an attempt to counter such selfishness, fines levied by the courts can be designated for conservation causes and smugglers' sentences can include suspension of hunting and fishing rights. "Swanson was feeding the market, feeding the demand," Oesterle tells me. "He was trying to monetize his hobby and doing what he could to skirt the rules." But Swanson wasn't involved in a big-bucks smuggling ring. The prosecution was motivated, at least in part, because of U.S. commitments to international animal protection agreements. "We have to fulfill those obligations," says Oesterle, "even though some box turtles and wood turtles may not capture the court's attention." Despite his vigorous prosecution of Swanson and the rest of the gang, Oesterle expresses no special affinity for turtles. "I'm not particularly a turtle lover," he allows, preferring to call himself a conservationist. "I'm not particularly a turtle guy, but I find them fascinating." He offers a personal note: "My son had one that walked away on us." Everybody has a turtle story.

"You didn't feel good about trafficking in turtles because you were engaged in a nefarious act or because your prized turtles were leaving home?" I ask Swanson at the IHOP.

"I don't know. Probably a little bit of everything. Plus they're in transit. I never heard if anything died. I'm assuming they did." His assumption is correct. Not only did Snaky write, "Yellow male lose one leg," but the indictment cites an email to Swanson from one of the Hong Kong–based conspirators, advising him that "one turtle had apparently died in transit." "There was a cost to the turtles," Oesterle confirms. "There was a loss factor that he was willing to accept in the shipment." Swanson tells me he knows it took at least a week for the turtles to arrive at their Asian destination, and "that's not acceptable for shipping a live animal."

I look across the pancake house table at Nate Swanson, a seemingly nice and mild-mannered all-American guy. "What was going on?" I ask him. "You knew you were breaking the law. You claim to love the animals, and you were jeopardizing the life of individual turtles while depleting species in jeopardy, sending them off to an unknown fate in Asia."

"I know." His voice is soft. "That's my struggle. I still love turtles. I still love frogs. At this point I probably love turtles more than frogs."

When the local sheriff, along with FWS agents, first banged on his door at the end of July 2012 ("Cop knock, you know? Super loud."), Swanson thought he was dreaming. As Swanson recounts the details of the raid, his mild-mannered tone changes and so does his vocabulary. The cops banged again. "I was like, 'What the fuck did I do?' 'Cause there hadn't been a warning." They had a warrant; he let them into the house and they searched. "They dumped stuff. They made a horrible mess. They found turtles." And they confiscated turtles, including black-breasted leaf turtles. Later they took others, including the Arakan forest turtle. As part of the plea bargain that resulted in his year-and-a-day sentence to prison, Swanson agreed to pay for the care and feeding of his former charges. Some went to the Woodland Park Zoo in Seattle, others to the Sarvey Wildlife Care Center in Arlington, Washington. The price tag for Swanson? A precise $28,583, which he agreed to pay off at the rate of $50 a month. Not that these caretakers' best efforts meant the turtles were safe and secure. In January 2012, the Sarvey Wildlife Care Center alerted the Snohomish County Sheriff's Office that turtles #12 and #16 were missing and the female eastern box turtle and the male three-toed box turtle were officially reported stolen from their new home in the center's volunteer lounge.

The eastern box and the three-toed were not necessarily Swanson's favorites. "You want what you can't have, what's unattainable," says Oesterle, trying to understand Swanson and his ilk. "But in this case they seem to love the species, they love the whatever it is they're collecting, but the mere collection is threatening to the species. So there's a disconnect." It's a disconnect

Swanson explains with a muscle car metaphor while claiming he does exercise restraint. "A lot of kids wanted a '69 Camaro," he remembers, "or a jacked-up Ford truck." Okay. "You get married and you have kids and the Camaro is the one thing you always wish you had." Right. "Well mine is a bog turtle."

The bog turtle (*Glyptemys muhlenbergii*) is the smallest North American species. It suffers from habitat loss and aggressive poaching for the pet trade. "I was offered bog turtles so many times and turned them down every single time," says Swanson, showing off his character. "I know their numbers in the wild. Certain things you don't fuck around with, and that was one for me. I would love to have one, love to. But I've heard of their mortality rates and of their superdeclining numbers in the wild. I don't want it," he declares. "I want them to stay put. Put them back. Don't talk to me." He's talking as if a poacher is at the IHOP table with us offering a deal. "I'm never dealing with you again. I talked to somebody on the phone who says they have ten of them. I'm like, 'What the fuck? How did you get ten?' From Pennsylvania of all places." Pennsylvania lists the bog turtle as not just threatened but endangered.[7] Offered as an example of his personal ethics and character, Swanson tells me he made a trip to South Carolina and witnessed an active thriving population of bog turtles, a sight that convinced him he could rationalize buying and selling that species. "I felt better about it."

Nate Swanson served seven and a half months in prison. "Everybody laughed when they heard I was in for turtles." (When I tell Oesterle the man he convicted was mocked by other inmates, he laughs too, agreeing, "I'm sure he was." Turtle smuggling, Oesterle imagines, especially as practiced by Swanson and his cohorts, is viewed in prison as a less than honorable crime by those convicted of what they consider more sophisticated frauds.) Swanson was sent to a halfway house for a spell and then freed on three years' probation. He came out "humbled," he says, and looking at things "more clearly." And the errors of his ways? Does he ruminate about his turtle mistakes? "I think about it every day. I walk by my turtles," not all of them were confiscated, "and I think to myself, 'Your herd was a lot bigger.'" He reminisces about feeding them discarded prawns from Costco. "I'd have a bag of forty-eight prawns and I would take all the shells off and these turtles would come running. They knew. That was a magical time for me. That was a big deal." It's not a bog turtle, but even feeding time for that herd sounds like a bigger deal for Swanson than a '69 Camaro or a Ford truck. "It felt like they were my children, not just pets. They knew I was coming. Yeah."

— *Busted in Louisiana, Locked Up in West Virginia* —

Convicted turtle smuggler Keith Cantore's home away from home reminds me of a community college campus. Poised on rolling bluffs along the east bank of the Monongahela River in Morgantown, West Virginia, the minimum-security prison looks bucolic on the lovely spring morning I visit. Single-story brick buildings dot the landscape as the incarcerated stroll the grounds—or at least their walks appear as strolls to this casual observer. Their khaki uniforms look like work clothes— no orange jumpsuits or black-and-white stripes here. No walls on the perimeter—not even fences. No locked gates. No watch-towers.

No recording equipment, either. "This is a pencil-and-pad in-terview, understand?" the guard who made the arrangements for me to spend a couple hours with Cantore reiterated when I showed up at the prison office with a yellow legal pad and an extra pen. My driver's license was all I needed to enter the visit-ing room, a place that looked more like a high school lunch-room than the Big House. No bulletproof glass separating me from the inmate, no telephone hookup to connect me with him—just a utility table and the guard within earshot on a raised platform off in the corner of the room. "Okay if we shake hands?" I call out to her. She shrugs her approval.

The Federal Correctional Institution, Morgantown, as the place is officially known, houses white-collar and nonviolent criminals, such as turtle aficionado Cantore. The (former, of course) mayor of Charlotte, North Carolina, was a resident, serving time for taking bribes. An Arizona congressman racked up convictions for extortion, bribery, insurance fraud, money laundering, and racketeering to earn his Morgantown address, and an Ohio congressman secured his for conspiracy to commit lobbying fraud. An inside trader received his Morgantown time for raking in almost $40 million in ill-gotten Wall Street gains. A *Survivor* contestant was in for evading taxes on the million bucks he won on the CBS television program. And Keith Cantore smuggled turtles, a violation of the Lacey Act of 1900, the first U.S. conservation legislation—the law that makes it illegal to ship protected wildlife out of its state of origin without a license.

We sit across the table from each other in that prison utility room, and Cantore tells me he's been catching turtles since he was eleven years old, growing up on the South Side of Chicago. His first turtle hunting ground was in his neighborhood, the Chapel Hill Cemetery. He found painted turtles (*Chrysemys picta*) and snappers. "I kept 'em as pets and I sold some to pet shops." He deploys an impish smile. "I didn't know then it was illegal to sell wild-caught in Illinois."

But a couple of years later, when he was thirteen, Cantore learned about what he calls "a flea market for animals" a little over an hour's drive across Chicago in the suburb of Schaumburg. There he developed a further affinity for all reptiles, but turtles and tortoises remained his first and true love. He says he shadowed veterinarians and copied them, learning how to diagnose ailments and treat them. He trained himself, he tells me, to conduct minor surgery, cure infections, and otherwise medicate his charges.

So you're something of a turtle scholar, I suggest.

"I've never been one to call myself an expert," he says. "I always thought it was kind of nerdy to get into the scientific names. I was more into the care of the animal and what I like about a species and how to make their surroundings comfortable for them."

There are no pauses when Cantore speaks; he talks as if there are no spaces between words and sentences. He sounds like a machine gun on automatic fire, with words coming out of him nonstop: "I've-never-been-one-to-call-myself-an-expert." He's been speaking so rapidly it's difficult for me to keep up with my note taking. Slow down, I request.

He stops talking for a moment. "I have nothing but time," he smiles and adds a sardonic line that sounds to me like one he's probably used before: "literally two and a half more years to be exact."

A couple of years before we meet each other in the prison, Cantore sent a message to one Lawrence Treigle, a Louisiana turtle merchant doing shady business supplying American turtles to buyers in Hong Kong. Cantore encountered Treigle's operation on Facebook and wanted in on the action. He dropped the name of a mutual acquaintance to establish his bona fides. "If I had just googled his name," he tells me wistfully, "I would have seen he had been arrested and made no deal with him."

But he did not bother with such due diligence and instead solicited a business relationship. "I buy all morphs and oddballs and wholesale stock for my Chinese customers," Cantore wrote to Treigle. "Let me know what you have or can get me." According to the federal agents who were eavesdropping, Cantore told Treigle he would buy all his wood turtles, paying $1,000 a pair. But that's not all. "I want a bog pair damn it," Cantore messaged. Cantore probably was familiar with the scientific name, *Glyptemys muhlenbergii*, even if he considers it nerdy, since bog turtles are among the rarest of North American turtle species, listed as threatened under the Endangered Species Act.

"Sweet," answered Treigle. "I told my buddy that we had another buyer that will take all the woods he can catch. He said its [*sic*] slow right now but when it starts to cool off he will get you as many as you can handle."

"Okay. Cool. Nice," responded Cantore.

A few days later Cantore and Treigle switched from Facebook to text messages, and Treigle offered to sell Cantore a pair of wood turtles to test their relationship.

"My buddy in PA is shipping me some tomorrow," Treigle wrote. "If you want I'll sell you an adult pair to start. Make sure everything is cool with you. I don't want to get my ass busted. I'm only trusting cuz of Mitch. If things go well we will work out more in the future. Cool? U not bullshitting me are u?"

But it was Lawrence Treigle who was doing the "bullshitting." At the time he and Cantore were exchanging these messages, Treigle was working for federal investigators, hoping to stay out of prison in return for trapping other smugglers. A month before this text-message exchange, a task force from the U.S. Fish and Wildlife Service, the Postal Inspection Service, and Homeland Security had shown up at the door of Treigle's home in Covington, Louisiana, with a search warrant. The feds were suspicious: $200,000 in wire transfers from Hong Kong to Treigle piqued their curiosity, along with boxes coming and going from his otherwise staid-looking four-bedroom brick midcentury contemporary house (with a pond in the yard)—boxes that the cops figured were packed with turtles.

Treigle sang. He explained the details of his smuggling operation to the agents. He acknowledged that he was trafficking in turtles without a Louisiana license for such trade and that the wood turtles he was peddling were caught in the wilds of Pennsylvania, another crime. Once he transported those turtles across state lines, Treigle was violating the federal Lacey Act. Read the comprehensive Lacey Act and you may hesitate to picnic on the California side of Lake Tahoe if your hotel is on the Nevada side, on the chance a protected critter hides in your picnic basket. Treigle didn't just sing; he signed up with the feds to sting his new buddy Cantore in hopes of avoiding prison himself.

"U want the 4 pair?" Treigle asked Cantore about the Pennsylvania wood turtles, the feds watching as he tapped out the text message.

"I want them now," Cantore wrote back, offering $900 for each pair. Cantore sent the money south, and federal agents packed four female and four male wood turtles into a shipping container and sent the box off to Cantore in Illinois. There a Postal Inspection Service agent—disguised as a mailman— delivered the package to Cantore's home. He came to the front door himself and signed for the turtles. The stage was set for the feds to up the ante of the sting operation. Cantore agreed to buy fifty pairs of poached woods for $40,000 and come to Louisiana to fetch them himself.

On September 5, 2014, according to the criminal complaint filed against Keith Cantore in U.S. District Court in St. Tammany Parish, Cantore sent a text to Treigle instructing him to meet at the Americas Best Value Inn and Suites on Gause Boulevard in Slidell ("Weekly Rates Jacuzzi Rooms," promised the marquee on the motel's signpost). Treigle arrived on the scene, his body wired for sound and driving an undercover police pickup truck packed with audio and video recorders. Thirty-seven wood turtles were secured in the pickup's bed. A nervous Cantore called Treigle and changed the meeting place to a Mc-Donald's a block down Gause Boulevard. At the restaurant Cantore instructed Treigle to drive home. He said he would follow in the BMW he was driving, and he told Treigle that they would do the deal at Treigle's house. He never made it. A St. Tammany Parish sheriff's deputy pulled Cantore over for, in the quaint language of the complaint, a stop "predicated on a traffic violation." In the back of the SUV, the cops found $20,000 in cash and ten boxes filled with turtles.

That's not all the government had racked up against Cantore as proof he was illegally buying and selling a substantive number of turtles and exporting them out of the United States. Evidence against him included Facebook screenshots captured by Treigle during the Slidell sting. A particularly engaging scene shows Cantore in a kitchen, the counter piled high with plastic tubs apparently full of turtles. He's staring out a sliding glass door, and it looks as if he is cupping a cigarette in his left hand. The telling caption reads, "You know the turtle hustle is seriously extreme when your kitchen has been takin [*sic*] over once again! And your turtle helper has to smoke a joint to cope with the hustle!!" Another snapshot shows a dish full of turtles and the caption is an advertisement: "Red eye Albino red ear group for sale 6k for group ready for export over seas!" Still another post publicizes what the ad calls "Flawless Grade A+ Albinos" including a lemon with red eyes, and it offers "32 red ears" for $16,000.

Keith Cantore pleaded guilty to violating the Lacey Act provisions that make it "unlawful to transport, sell, receive, acquire, or purchase in interstate commerce any fish or wildlife taken, possessed, transported, or sold in violation of any law or regulation of any State or Foreign law or attempt to do so." Both Pennsylvania and Louisiana laws prohibit catching, selling, and shipping turtles without a license. But this was not Cantore's first violation of the Lacey Act. Back in 2008, he was convicted of selling undersize turtles, and his police record includes gun and drug convictions. Those priors didn't help him at sentencing time after he plea-bargained the 2014 Slidell turtle charges against him down to one count. His penalty: forty-one months in prison, plus three years probation, plus almost $43,000 in restitution charges.

"The time don't really teach you a lesson," he tells me in the Morgantown visiting room. "I got more time for this than I did for machine guns in Chicago."

Machine guns?

A slight smile from Cantore as he reports, "In Chicago you kind of need them. Before I got into turtles, I was known as a gunrunner. I was a little bit of a wild child." But he still cultivates that bad-boy image, as his marijuana-decorated Facebook page bragging shows. Of course when Cantore talks about running guns before he got into turtles, he's talking about running guns before he got into the *illicit* turtle trade. He wasn't running tommy guns when he was an eleven-year-old kid in Chicago chasing turtles around the neighborhood cemetery. And it's turtles all the way down for him, despite his felonious setbacks. "I'm absolutely not leaving turtles," he says as the guard listens. "But when I get out of here I'm going to try to do everything legal. In this case," he says about the Slidell bust, "I didn't know the turtles were endangered," he leans back in his chair and adds, "but I knew you weren't supposed to sell wild caught."

It's understandable that Cantore wanted bog turtles. "The bog turtle undoubtedly is the rarest and most valuable native American turtle in the trade," Eric Goode says when I check in with him about the black market value of the animal topping Cantore's wish list. "It's been the rarest turtle in the United States for a very long time, rare, sought after, and coveted." Only about four inches across its back, the bog is also one of the smallest North American turtles, says Goode, and elusive, because finding the sphagnum bogs where the turtles like to hang out usually requires more than a casual Sunday drive through the countryside. Successful bog turtle poaching usually means that experienced local guides serve the hunt. Eastern Pennsylvania offers prime bog turtle hunting grounds. "If your jailhouse subject sought this turtle," Goode tells me, "he must really have known what he was doing."

Not that those bog turtles Cantore chased would necessarily have been destined for the Asian market. "Bog turtles don't survive well in China," James Liu told me. Liu manages the Turtle Conservancy's Asia programs, and he had just returned from a trip to the Shunde Turtle Breeders Association exposition and conference in Guangdong Province, where he encountered no bog turtles on display or for sale. "They are a novelty because they are so hard to get." But at the expo, Liu met turtle buyers who dismissed bogs with explanations such as, "Yeah, I bought a few of those and they all died. They're too expensive. They don't seem like they're worth the money." Perhaps their perishability will protect the bogs from the insatiable Asian turtle market. "Why spend thousands upon thousands of dollars to buy this relatively unimpressive small turtle," figured Liu, "a turtle that's probably going to die in Asia, especially in the hot climate of southern China."

Keith Cantore may have been seeking bog turtles for their fat profit potential on the American black market, but morph alligator snapping turtles, those unexpected oddities in coloration that show up in hatchlings intentionally bred for such surprises, are Cantore's personal favorites. He looks at me with what I'm fast learning is an engaging smile that he easily turns on and off, this time at the memory of his collection of designer turtles: albinos, pinks, whites—he estimates fifty or sixty morphs were seized by authorities after he was arrested, including a silver and a cinnamon. "I had the largest collection in the world," he boasts and laments at the same time, insisting such a massive pile of the fierce-looking turtles was legal in Illinois when he assembled his collection. Alligator snapping turtles are listed as endangered by the Illinois Department of Natural Resources, which warns in its literature—in uppercase letters— that "harvest of alligator snapping turtles is ILLEGAL in Illinois," and has been since the end of 2014.

"I love snapping turtles," Cantore says, becoming more and more animated—gesticulating with his hands—a love that apparently transcends whether they are pets or merchandise (or both). "Certain turtle species are endangered," he agrees, "but you'd have to be a fool to think alligator snappers are endangered." He reminisces about a couple he described as over five feet across the carapace. "I paid $7,000 for the pair. I don't know why I like them." But he does offer a clue. "I like mean or vicious animals. I feel bad for them because people don't like them. Alligator snapping turtles, crocodiles, mean dogs."

Cantore leans back in his prison visiting room chair again and considers further his affinity for the mean and the vicious; he smiles as he finds a clue in his early run-ins with the police. "I'm an adrenaline junkie. I used to think it was fun getting chased by the cops." Chase him they did. In addition to his prior convictions on gun and drug charges, the government claimed Cantore continued to sell turtles illegally while he was out on bond as the Slidell charges were being adjudicated. In his sentencing memorandum to the court, the U.S. attorney on the case, New Orleans native Kenneth Allen Polite Jr., noted that after his Slidell arrest, "Cantore continued to show his disregard for wildlife protection laws by illegally selling turtles [making] numerous postings to Facebook and Fauna Classifieds, an exotic animal website, offering turtles for sale." The government pointed to Cantore's criminal record as a rationale for a stiff sentence. "Cantore's prior convictions not only have failed to deter his criminal behavior," stated the memo, "but also have served as a means to further that criminal conduct." Polite points to Cantore's claim to Treigle that he could be trusted precisely because he had served time for weapons trafficking and selling turtles illegal to trade. "The lesson Cantore learned from his prior convictions is that they prove criminal bona fides, not that they represented a lapse in judgment that would not occur again."

Only a month before Cantore was sentenced for the Slidell caper, he was picked up in Will County, Illinois, and charged with possessing or offering for sale two alligator snappers along with six alligators and fifty-four common snapping turtles. The charge: he held no valid aquatic life dealer's license, as required by Illinois law.

Why those lapses in judgment? Now I'm the one leaning across the prison table as I ask Cantore about his motive, figuring it must be the bigger and bigger bucks now attracted to the turtle business, legal and illegal. "I guess the money," he allows, laughs, and then quickly tries to put his answer into a rationalizing context. "I didn't get into it for the money. I really don't do it for the money." The tough-guy face looks boyish for a moment. "I like turtles. Turtles used to keep me out of trouble," he suggests, "until I was offered the bite." After he was released from prison in 2009, he claims he went straight. "Then I got the offer," he says about the sting deal. "It was a chunk of change. And here I am." Based on the government's case against him, of course, he's skipping over some germane details about his turtle trading antics after that first stint in federal prison. But he goes right for the punch line. "If I had made it home that weekend," he says about the trip to Slidell, "I would have made $100,000." He looks at me, and he smiles yet again. "So now you know why people do this. One week of partying and I make more than most people do in two or three years."

The risks attached to trying to score a hundred grand with turtles rather than AK-47s or cocaine may be another lure for dealing turtles. It's hard to imagine that a conviction involving guns or drugs worth that much cash would result in such relatively light prison sentences as Cantore and other animal smugglers often receive. But Cantore wants me to believe that it's not just his risk analysis for illegal animal trading that led him to trafficking turtles. He genuinely enjoys their company. "They're interesting animals. Turtles interact with you when you feed them. It's fun watching. They definitely know you. If I walk into the room," he says about his turtle stash points, "all the turtles go nuts because they think I'm going to feed them. They come to me, just like your dog." His tortoises, an African sulcata and an aldabra, he insists "come if you call them by name."

But what is it about the combination of guns, drugs, and

animals that lures the criminal mind? It is a threesome I frequently encountered while researching the lure of exotic pets—like big cats, long snakes, and great apes. The guns-and-drugs symbiosis is obvious: the same transit corridors and a nefarious clientele with a propensity to seek both. Why add animals to the mix? It's not only because an animal-trafficking rap usually means less time behind bars than those for guns and drugs. It's not just a random ripening of a childhood hobby, such as Cantore's and Swanson's apparent love for turtles. There is plenty of money to be made trading animals. But there's another factor. Fetishists often seek live animals simply as one more selfish acquisition. Throughout history humans have sought other animals for their own entertainment. Under the best of circumstances, the feelings are mutual and the animals thrive. But more often the relationship is one of the humans subjugating the animals. Reducing animals to playthings—objectifying them—can be a cruel as well as an unusual pastime, but it's also compelling for some types. *Jefes* in the Mexican drug trade, for example, are famous for their menageries. "It's a show of power and is incredibly common in the criminal underworld," said Patricio Patrón Laviada, when he served as Mexico's attorney general for environmental protection. "The worst of the worst have exotic animals."[8]

"It's got to be the rush and the money," figures Cantore, along with his love of the product he's smuggling. Because he served as a peddler, his suppliers often offered him free animals for his own use, "just like drugs." Once initiated into the world of exotic animals, he tells me, it's easy to slip from the legal to the illegal. "When you sell legal stuff to Asians, they'll ask you for something 'special' that's illegal. That's what sucks you in. I stayed legal until 2014," says Cantore. He tells me Facebook was an ideal tool for him to "reel in the buyers," buyers looking for turtles illegal to possess. "They hunt me down on Facebook," he says, based on his seductive advertising pictures and captions, with messages such as, "I know you can get them." His responses, he says, would be coy lines back like, "I don't mess with those," to which the buyers would urge back, "Oh, I know you could connect me." After a few more such exchanges, complete with fake names, enough Facebook-fueled trust would be established to rationalize exchanging mailing addresses, money, and turtles. "Over there," Cantore says about the appetite for all things turtle in Asia, "it's almost like a religious thing. That's what my Asian friends tell me. It's almost like they worship them." He pauses and reconsiders. "Not all turtles. The cheap ones they eat." Neither his sweeping analysis of what he perceives as the rationale for Asian obsessions with turtles nor his explanations of turtle physiology stop. "Turtles' and tortoises' internal organs don't deteriorate. Asians are infatuated with that fact. That's why Asians eat them."

We're nearing the end of the two hours allotted by the prison for our visit. Cantore still is speaking as fast as Aesop's hare runs. I ask him about his suppliers, the poachers who catch turtles for him in the swamps of Louisiana and the bogs of Pennsylvania. He's brutally dismissive, saying they're often so ignorant about animal husbandry that the turtles' lives are in jeopardy. "The majority of these assholes don't know and don't care," he says, calling them "hillbilly dumb-asses." He cites turtles he's seen advertised online that appear in publicity photographs obviously not well cared for. "Smugglers usually are bad with the animals," he says, while applauding himself for bringing to his home turtles that look abused. "I'd buy them and take care of them. It was illegal to buy them, but I wanted to see them out of those conditions. I could tell from the pictures that the turtles were not eating. They starve, piled up in buckets and Walmart plastic totes until they sell." He's not only talking fast but also reinforcing his points with aggressive hand gestures. If no farm is listed online as a source for turtles on offer he considers it a sign the animals were caught in the wild.

Cantore is an engaging storyteller and a well-spoken man, his passionate speeches about his love for turtles marked only rarely with grammar errors. Not as rare are his crude, aggressive rants against poachers and fellow smugglers. "There's a huge number of 'em been in prison for drugs and guns. There's some fucking nutcases out there. Drug addicts. Ignorant assholes," he calls traders showing their wares online via photos of ill-treated turtles, and then he considers his own credibility. "I'm not one to judge because I broke the law." Indeed, and who besides Cantore knows how many of those rescues he says he saved ended up as revenue generators for his illicit business.

"Prison doesn't bother me," Cantore tells me as I'm about to leave the penitentiary. "So sending me to prison is probably the wrong thing to do to me. I'm not scared of prison. I've got no wife and no kids. It's just a time-out for me. Time to get back in shape." In fact he does look more fit in person than his mug shot image, a picture that shows him thick-necked, his dark brown hair chopped short and his eyes emotionless—a face with no smile, no frown.

What should his penalty be if not prison? "I hate fines," exclaims the capitalist smuggler, "and community service sucks. Make me breed animals, do something useful. Why screw the taxpayers out of hundreds of thousands of dollars on my case and on prison time? I'm giving nothing back to society or to the animals here."

When he finishes serving his sentence and walks out of the Morgantown penitentiary Keith Cantore expects to head back to his Illinois home and join his family's air-conditioning business. But he'll still be dealing in turtles. "I have two-headeds and Siamese in the freezer," he says longingly. "Once I get out I'm going to see if I can reproduce them and sell them to freak shows." He's excited remembering the shape of the shell of one Siamese that "looks like a heart. It has six legs and walks like a crab!" Such unexpected oddities, he says, make the best acts for shows. He announces again that he's finished with smuggling, even though he acknowledges it's fun. "It's an adrenaline rush, just like any crime," he displays an impish grin. "But I can't do it again. I can't come back a third time. The judge will hammer me." Before we say good-bye, Cantore insists, "I'm absolutely not leaving turtles," and adds for the record (and probably for the guard listening to us), "but I'm going to try to do everything legal." Intriguing (and perhaps realistic) his decision to add the not-so-subtle loaded word "try."

After Keith Cantore was sentenced, Immigration and Customs Enforcement Special Agent Raymond Parmer, whose agency helped with the investigation, expressed disgust for wildlife smugglers. "The illicit trade in threatened and endangered species represents the destructive results of unfettered greed," he announced, and then fingered Cantore as an example. In effect, "this defendant was willing to help drive wood turtles to extinction to make a few bucks."[9]

Wildlife cops, vastly outnumbered by criminals preying on the easy targets of susceptible animals, work against the odds to enforce the law and protect turtles.

• Sloppy Fred •

Breakfast time for Fred. He's been climbing the walls (literally) and giving his empty food dish forlorn stares (when he's not sitting in it). I cut up some apple and mash up some leftover salmon, lift up the grate on the top of Fred's house and dump the mixture on his dish. He's hanging out right next to it and just after the food hits he pivots, looking like an athlete in a slo-mo replay. His deliberate movements rotate his body until his head is in position right over the breakfast, and then the grub gets his attention, 100 percent. His neck stretches, his beak pecks, as he makes it clear he prefers the salmon this morning over the apple.

My own breakfast is getting cold while I continue to watch Fred chow down on his. His actions are slow and steady, except when he exercises the crucial grab. Why the allure? It's just a turtle eating leftovers from last night's dinner. Maybe it's the primal nature of Fred's connection across eons to the dinosaurs. Maybe it's because I feel some responsibility and affection for him, and I'm happy he's eating, because it means he's healthy (and maybe happy, too). Maybe Fred's attention to detail is a lesson in mindfulness, a lesson about being present and living in the moment.

As I leave Fred to continue my morning routine, he's half-sprawled in his food dish, salmon slopped all over his face.

EIGHT

THE FRUSTRATED COPS

— Stalking Turtle Boots —

The scene is a McDonald's on Redondo Beach Boulevard in Gardena, the heart of the Los Angeles sprawl. Juan Ramirez, an inspector with the U.S. Fish and Wildlife Service, is stripped down to his undershirt as Special Agent Ed Newcomer straps a transmitter to his chest. It's Ramirez's second encounter with a shopkeeper who sells *ropa vaquera* cowboy clothing—out of a mom-and-pop shop on New Hampshire Street.

Inspector Ramirez stumbled on the cowboy duds store one day after lunch and decided to check out what was on offer. While Ramirez was looking at boots, the shopkeeper made a suggestion. How would he like a pair of "special boots"? The conversation was in Spanish. Inspector Ramirez was offered cowboy boots made of *caguama* for $750. Any trade in anything made from sea turtles is illegal.

"He's just going in today to howdy-and-shake," Agent Newcomer says. "We're not going to make a buy. He's just building on that relationship he started last week." Traffic on Redondo Beach Boulevard is competing with our conversation, as are birds singing in the trees. Los Angeles is a clash of the hyperurban and the resilient wildlife.

Newcomer is finishing up with the covert equipment. "This

is the transmitter." He points out a unit about the size of a mobile phone. Whatever transpires in the store will be broadcast
to Newcomer's car, half a block away, where he'll be listening
and recording. The wire is not just to gather evidence. "If something were to happen, Juan would say something in English
that would cue us that it's time for us to come rescue him."

"Does that make you uncomfortable?" I ask Ramirez. "Are you
nervous about going back in there knowing you're the good guy
and they're the bad guys?"

"No," he brushes off the idea. "I flow with the conversation.
I'm a customer. That's all I think about. I'm just in there to buy
something."

"If something were to happen," Newcomer explains, "Juan
would say something to the effect of, 'Hey, you don't need a gun,'
or 'Don't point a knife,' or 'I'm no threat to you.' If we hear those
words, he knows we're going to be about twenty to thirty seconds
away."

— Wildlife Forensics —

Just a few hours' drive from my university campus through the
verdant southern Oregon hills on Interstate 5 is the California
border city of Ashland, famous for its Shakespeare plays and
little known for its U.S. Fish and Wildlife Service Forensics Laboratory. The FWS notes that this is the only crime lab in the
world devoted solely to the enforcement of wildlife laws. The
primary role of its scientists is, according to its mission state-
ment, to "link suspect, 'victim' and crime scene through the
examination of physical evidence; determine the cause of death
of wildlife crime victims; and help analyze crime scenes." The
word "victim" is in quote marks because the lab finds itself speaking for victims who cannot testify for themselves: wildlife.

In a nondescript building on the outskirts of Ashland I find the lab's herpetology specialist Barry Baker, an archaeozoologist by training, a scientist who studies the use and abuse of animal parts by humans over the last ten thousand years. "I love turtles," he tells me as we sit down in his office, an office littered with animal memorabilia, including plenty of turtle tchotchkes. "I have many turtle stories," he says. Smiles come fast and easy from the gray-bearded Baker as he recounts some of the better ones to me from behind his desk, acting as relaxed as he looks in his blue jeans and red-and-white-plaid short-sleeved button-down-collar shirt.

— Analyzing Turtle Lure —

"They're slow," Baker muses about why turtles hold such appeal across cultures and eras. "They're easy to catch and relatively easy to keep as a kid, a nice nonthreatening pet if you're venturing beyond a cat and a dog." But there's more: turtle mythology through the ages. "There's a huge mythological aspect," he agrees. "The fact that it has a shell, that it can retract. We have those ancient stories of turtles and tortoises—it's turtles all the way down in cosmology. They're just such a different, unique group of animals, different from anything else." That may be at least an origin of the lure they hold. *Vive la différence.*

Turtles evolved structural differences over time. Today's models are not exactly what was roaming Earth in the time of dinosaurs, but they're of the same ilk—sort of like how both my retired 544 and current S-60 are both Volvos even though fifty years separate their design. I can squint looking at the two and see that they're related.

Dallas was Barry Baker's childhood hometown. He grew up in a neighborhood where he could find wild turtles. "I caught a box turtle as a young kid and kept it in the garage." It featured yellow patterns "like fireworks," so Baker named it Sparky. "I had him at least a year or two and then I let him go back in my yard. He disappeared. A few years later I saw him again in the same yard."

That love of turtles motivates his forensics work against the bad guys devastating turtle populations worldwide. He identifies animal remains. "Knowing what the animal is," he explains, "is important for knowing if a crime has been committed." He looks at turtle bones and shells, and artifacts made out of turtles, determining if they once were part of a protected species. Baker can take a pair of cowboy boots and determine if they are made of sea turtle. He can study guitar picks (and eyeglass frames) and know if they are faux or real tortoiseshell. From a sample of a hunk of meat he can determine if a restaurant— such as the one where I ate in Havana and the waiter offered me *tortuga* for dinner—really is serving turtle or is hustling the *turistas* with something mock.

"There are so many different species and they're encountered by people all over the world," Baker says about turtles and tortoises when I seek his theories as an archaeologist and an anthropologist about why the animals captivate humans across time and cultures. "It's not just like a rhinoceros, which you may find in Asia or Africa. You can find turtles everywhere." In the sea and on land, from deserts to forests, turtles are everywhere. "Turtles are somewhat secretive. Sometimes you stumble across them, but sometimes you have to look for them," Baker notes as he speculates about their intrigue. "They're tough. They're very resilient. They can protect themselves with their hard shells. There's something very charming about turtles," he says, perhaps bordering on anthropomorphizing the reptiles when he adds, "and some of them have personalities."

"Baloney!" I interrupt, because I'm a cynical reporter and maybe also because I wish Fred would express some emotion toward me besides curiosity.

But the scientist shouts down the journalist, insisting, "No, they do!" He, too, insists—as did Davey Evans at the Concordia

Turtle Farm in Louisiana, along with so many other turtles aficionados I've been talking with—that turtles can recognize individual humans. "There's an ancient connection with them,"
he theorizes. Baker recounts meeting Galápagos tortoises at the
San Diego Zoo. "It was an overwhelming feeling of connecting
with something ancient. I think that resonates with a lot of
people."

— *Fake Tortoiseshell ID* —

Barry Baker takes me behind the scenes of his lab for a lesson.
How, paraphrasing Cole Porter, to differentiate between artifacts made from threatened, protected sea turtle shell and the
mock—those made with plastic. He shows me a barrette (it's a
sweet little number, decorated with a floral inlay) and proceeds
with his standard investigative protocols. "I'm going to look at
it visually," he says, "to see if it looks like it's real or fake." He
checks the color and notes amber and some dark brown. Then
he points out striations that he theorizes are cracks resulting
from stress during the bending of the barrette material when it
was fabricated. He holds the barrette up to a light and shows me
how the dark brown bleeds into the amber. He notes a lack of
symmetry. He spots scars that make it look to him carved and
cut, not molded in a plastics form. "All those clues tell me that
this is from a hawksbill sea turtle," from the scutes of the shell.
"I can see tool marks," his visual detective work adds, "from
where it's been cut or sawed." Next step is to show the barrette
to his chemist colleagues. If it were obviously plastic, he would
not bother them. "But if it does look real to me, as a second step
of verification, I would give it to chemistry, they would sand it
and analyze that dust for its chemical signature."

I wonder if Baker is disturbed and distraught looking at a
fashion decoration, knowing it's fabricated out of an endangered
species. Or is he inured to such damage after seeing so much of

it. "It depends," he admits. "It depends on what day it is. It depends on how much of it I see. Some days it is very overwhelming. Especially if I have the whole dead turtle." He points to the barrette and says he understands how something so pretty can be sought for vanity and bought without qualms. "Many people don't realize it comes from a turtle even though it is called tortoiseshell. They just assume it's plastic." Or if they suspect that it is an animal product, they think it was part of a tortoise because of the term "tortoiseshell." "They don't think sea turtle." As Baker talks, his mood changes from science to emotion. "Yeah," he allows, "it does hit me. I can be overwhelmed because day after day after day, it is death after death after death." Advances in science that develop from the forensics work keep Baker and his colleagues in the lab sane, he says. "We live for that."

What also helps is a growing public fascination with sea turtles. A swim with turtles is showing up on plenty of bucket lists. "An awesome sight when one swims past you in the water," is the enthusiastic call from the Lonely Planet guidebook to Maui. "There's something deep and mysterious about them in terms of time and oceans," Barry Baker says about sea turtles. "They migrate. They disappear and they come back. Somehow they know how to come back, and when they come back they often come back en masse." But scientist Baker doesn't shy from adding the word "cute" to his analysis of why any type of turtle can attract human interest. "There's something cute about little baby turtles. A lot of people remember the dime store turtle in the plastic container with the palm tree." Ah, speaking of iconic. "I had one of those, too," he laughs at the memory.

Exhibit B on the lab table is an elegant cowboy boot, a boot Baker proves is made with leather harvested from the legs and throat of a sea turtle. He shows me a taxidermied turtle and compares its scales with the shapes of those on the boot—scales unique in the natural world to sea turtles. Counterfeit-turtle boot makers stamp cow leather to make it look like turtle, but telltale evidence of hair follicles gives their fraud away; this boot has no hair follicles. He augments his visual inspections with citations from natural history literature. "It's not unusual for me to be looking at a paper that was written in 1832 by somebody who described the hair of an elephant's tail," he says, pointing to the lab's library, and such obscure citations get included in the documentation Baker submits to the court. Science-based proofs are required for the courts, but Baker's trained eye knows what's real and what's not and it's an eye he can't turn off when he clocks out of the lab. "Drives my wife crazy when I'm looking at snake leather women's shoes as we're going through the department store and I'm saying, 'That's fake, that's real, that's fake!'"

Which makes me wonder about my classic pince-nez. The frames look like tortoiseshell to my untrained eye, and they look so cool with their brass clamp to hold the round James Joyce lenses on my nose. Real or fake? Barry Baker says he'll know once he inspects them. First, he offers an official definition. Tortoiseshell, as fashion terminology historically applied to the real McCoy, means the keratinous scutes of the hawksbill sea turtle. Those scutes are the sections of the shell that together form the hawksbill carapace, overlapping each other like roof shingles. Heating the scutes makes the plasticlike material pliable and easily formed into decorative inlay for furniture, barrettes, and eyeglass frames.

"I will look at them," says Baker, "and see if they have the visual characteristics of something that comes from an animal as opposed to being some type of plastic." Where the amber and brown pigments of true tortoiseshell merge, the dark bleeds into the amber. But on the fakes, says Baker, the colors contrast with a sharp edge, not a gradual permeation. "I'll look for bubbles. If it has bubbles in it, that tells me it's probably plastic." On artifacts made from real tortoiseshell, marks from carving tools are evident, as are residue stress marks from when the piece was heated and bent. Telltale giveaways for plastic are seam marks from molds. "But because there are increasingly good fakes, we take a little piece of sandpaper, sand it, and analyze that dust. We zap it with a laser." How the laser passes through that material gives a signal of its chemical composition. Still on the drawing board when Baker and I talk is DNA analysis to identify what comes from scutes and what's plastic.

— Fake —

A few weeks later my pince-nez comes back from the lab—my phony pince-nez. Archaeozoologist and forensic herpetologist Barry Baker called them fake as soon as he glanced at them and saw that the darker pigment was more reddish than sea turtle keratin. To confirm what he saw, he sanded the frames, collected the dust, and, using Fourier transform infrared spectroscopy, zapped the dust with the laser. "Materials absorb and disperse laser light at different wavelengths depending on their molecular-chemical structure," he explained. "The result is a spectrum-graph of what this reflection and absorption looks like. That spectrum from the dust of the unknown sample is then automatically compared by the software to the database of the known samples. We've created an in-house database over the years of what such a graph looks like for a diverse range of plastics, keratins, and other materials." My pince-nez matched plastic. I can wear them with a clear conscience and augment my eccentric professor act.

In addition to the plastic tortoiseshell pince-nez, Baker sent a note with further philosophical ruminating about the human condition vis-à-vis turtles. "Through time," the scientist wrote, "they have been worshiped, kept as lifelong pets, and used as cosmological proxies for understanding our very existence. Yet with the turn of a blade, we butcher them, eviscerate them, devour their flesh and eggs for small doses of protein, and use their parts as trinkets in an unsustainable global economy focused on financial profit. We are puzzled by the evolutionary origins of their iconic vertebrate shells; yet it is us, as humans, who are the very strange species indeed. How did we arrive at this place, from primordial seas, to become the less excellent fishe?" he asked.[1]

— *Shell (Oil) Protects Turtles* —

A twenty-four-hour layover in Paris is a divine interlude for a three-continent hop from the West Coast of North America to sub-Saharan Africa. Wandering around the Latin Quarter, strolling along the Seine, lighting a candle in Notre Dame, buying a cheap umbrella—even cold rain can't interfere with enjoying Paris, especially a Paris all but free of tourists because it's gray and still feels like winter. A Niçoise salad, a glass of red wine, and a coffee were fuel for the taxi ride back to Charles de Gaulle Airport and the Air France flight south to Libreville. Standing in the aisle, anxious to get off the plane after seven turbulent hours, a fellow passenger welcomed me to Gabon with a natural inquiry. "Do you speak French?" he asked. "It's not easy, like English." But that was not his punch line. He smiled and added, "Wait until you meet our mosquitoes!"

A few days and slathers of Deet later, it's a late springtime night with a half-moon high over the West African beach that's just bright enough to illuminate the tracks of sea turtles heading up out of the Atlantic. I'm trekking along the tide line with a trio of Gabonese workers, patrolling for turtles and their nests. We're equipped with tags for the mamas as we look for new nests. There's little worry about poachers here near Point Dick. The sea turtle refuge is adjacent to Shell's sprawling Gabon oil works at the ramshackle boom-and-bust city of Gamba (there's not much more oil to extract from these fields). But what Shell's military-like base of operations means for the turtles is that human predators can't get close to their eggs. Since oil was discovered in the Gamba patch back in 1967, only visitors authorized by Shell can gain access to this beach.

We walk fast in the deep sand. *"Hippopatame!"* Our troop's leader, Mboumba Brice, breaks the silence with an urgent whisper, holding out his palm to stop the posse. *"Une maman et le bébé!"* We four freeze while the huge beasts—the baby looks the size of a VW Beetle—lumber over the sand from the forest across our path and splash into the ocean, leaving their huge and deep tracks in the sand. Just minutes later another massive and familiar shape comes into soft focus through the gauze of the moonlit night: the silhouette of an elephant standing alone on the broad beach, our second elephant sighting that night. We give it wide berth, silently bypassing the mammoth creature. We encounter no turtles and no nests this night, but no complaints. Hippos and elephants ain't bad.

"Watch out for the elephants," I was told by my Gamba hosts, and elephant warning signs dot roads on the Shell grounds. "Give them the right of way and no flash photos. They might charge you and we don't want that." No, we don't.

Mboumba Brice works the beach barefoot, his jeans rolled up to keep them dry. A stylized image of a turtle decorates his T-shirt, the trademark of Ibonga—both the name of the turtle protection group he works for and the word for "turtle" in the local language, Baloumbou. Brice and his colleagues fence the turtle nests they find, hoping to keep natural predators such as monitor lizards from scavenging the egg clutches.

I get a call early in the morning at the house Shell provided for me during my stay at Gamba. A clutch is hatching at Point Dick. I jump in the pickup truck assigned to me so that I can drive myself around the Gamba Complex of Protected Areas and head for the beach, careful not to exceed the slow sixty-kilometers-an-hour speed limit. Shell equips its trucks—including this one, from the Smithsonian Institute's research center—with strict governors. If I go too fast, the Smithsonian gets a ticket. Too many speeding tickets and Smithsonian scientists lose their Shell authorization to drive on the oil company's property. Drive over seventy and the truck automatically shuts down, leaving rule breakers stranded who knows where on the complex with elephants and hippos and gorillas and leopards for company.

— *Close Encounter with Fred's Cousins* —

Out on the beach I switch into four-wheel drive, careening as if the sand is ice until I arrive at the nest in question. Mboumba Brice is on his knees at the nest, digging fast and carefully, pulling just-hatched olive ridleys out of the hole their mother dug when she picked this spot to lay her eggs. Brice is counting as he grabs the little turtles, each about the size of a silver dollar and looking to me like identical siblings. He pulls them out and puts them in a holding pen that suggests his sense of humor: it's a green plastic kiddie wading pool—in the shape of a turtle. I take one of the little fellows into my hands. Its flippers tickle, but it exhibits zero interest in me. Why should it? It just wants to spring along with all the others into its Atlantic home.

A few human mothers come up the beach with their children to witness the exodus from nest to ocean. They are Shell worker families living on the Gamba complex. As the youngsters marvel at the miracle, one of the mothers tells me it's common for Shell employee offspring to pursue professions dedicated to conserving natural resources and protecting wildlife. "I think it is because they are exposed to things like this," she says, as Brice allows the children to play with the hatchlings. "And it makes up for their parents' shortcomings," she laughs about careers in extraction industries. Another mother tries to explain why the little turtles fascinate. "They're adorable," she says. "All babies are adorable, it doesn't matter what they are."

I'm gentle, but the olive ridley in my palms struggles and scurries, trying to get free of me. I put him down. He doesn't look back. He (she?) heads for the surf, crosses the tide line, and races—almost harelike—into the water. My little guy gets hit by a wave and tumbles into somersaults. I can still see him when the next wave hits, sending him spinning like clothes in a dryer. Then he disappears westbound into the Atlantic to figure out for himself what to do next. "Alone, with no sense of being alone," D. H. Lawrence calls it in his poem "Baby Tortoise."[2]

A smiling Mboumba Brice looks pleased as the surviving turtles—over a hundred from the one nest—all disappear into the sea (a few eggs did not hatch and a couple of turtles died before he dug up the nest). "I've done my job," Brice says. "I hope nature will do the same." Those turtles that do survive coming of age in the open ocean—a small fraction of the clutch—will come back to Point Dick and mate. The females will flipper themselves up on the beach, dig nest holes and fill those nests with their own eggs, eggs with a better-than-most chance of survival because of their odd bedfellow: Shell Oil.

— *Bad Guys and Good Guys* —

Ah, Royal Dutch Shell—an ultimate extraction industry bad guy. Ubiquitous and convenient Shell filling stations line our highways, but the Shell reputation is not limited to roadside gasoline and toilets. Its name is linked by Amnesty International to alleged atrocities against the Ogoni tribe and their home on the Niger Delta. Shell calls the allegations "without merit." From Russia's Sakhalin Island to County Mayo in Ireland, from Alaska's Arctic to the Gulf of Mexico, environmental activists like Friends of the Earth target Shell as a villain—cutting down forests, polluting air and water with refinery debris, flooding sensitive lands and seas with oil spills. Nonetheless, the Smithsonian chose to make a deal with the oil company, taking Shell money to finance its research station. It's a relationship Shell touts in its Gabon propaganda. "In 2000," the company boasts, "Shell Gabon established a partnership with the Smithsonian Institution for a biodiversity monitoring program in the Gamba Protected Areas of Gabon."

Save for Shell's oil works, the Gamba Complex is wild and undeveloped. The biodiversity Shell pays the Smithsonian to monitor is extraordinary. "The area harbors astonishing wildlife, including elephants, gorillas, chimpanzees, four species of marine turtles, manatees, buffaloes and hippos," gushes the World Wildlife Fund, which operates outside of the Shell lands. "Just offshore, endangered humpback dolphins and their cousins, the bottlenose dolphins, can be seen swimming in the surf. Between the months of July and October, humpback whales can be observed from shore or from sea as they perform their annual migration to breeding grounds just off the coasts of Gabon."[3]

Shell proclaims that its partnership with the Smithsonian is not just a public relations project but is designed to pursue three specific goals: "Minimize the impacts of oil activities and other industries by integrating biodiversity conservation. Promote education, training and increase awareness on biodiversity in Gabon. Carry out conservation and sustainable development projects in the Gamba Protected Areas."

My host in Gamba, Lisa Korte, the Smithsonian's Gabon station chief, agrees. "The Shell concessions," she tells me in her delightfully air-conditioned office, "are basically better protected than the national parks." That's because Shell's operations are guarded like a military base. A fringe benefit of protecting the oil operations is a safe harbor for wildlife.

Dr. Korte is a conservation biologist by training. Elephants are one of her specialties. The elephant warning signs on the Gamba roads offer a daily reminder that her subjects surround her. Her smile as we talk is a wide one; she likes her work and she likes where she's posted. A big picture of an elephant in the bush adorns her office. But her experiences in Gabon led to another expertise: a study of how extraction industries can work with conservationists to secure biodiversity in places like the Gamba oil fields.

Of course I ask if the Smithsonian contract is a deal with the devil.

Nope, she says, and insists she's sanguine about the relationship. Shell pays the bills for the research the two entities agree Smithsonian scientists will pursue, work "that we do in cooperation with Shell, but that we do independently of Shell." Shell cannot censor the Smithsonian, says Korte. "We own all of our data and we have the right to publish it, so that helps us maintain credibility and a certain amount of freedom and space from Shell Gabon." Crucial is the phrase "a certain amount of freedom." Not only does Shell underwrite the Smithsonian's Gabon budget, the company provides access, trucks, office space, and Korte's comfortable home—a tract house that looks like a Florida bungalow. "We're embedded in the company." She and her staff agree to abide by Shell's rules, just as a reporter embedded with the military must follow orders or lose access to the troops. "We're working closely with them and that gives us advantages. It helps us be eyes on the ground and we can make changes, we can offer advice. It doesn't necessarily mean the company will take it, but at least we have the ability to communicate with them." Besides, she says, the environmental problems throughout Gabon—from trash in the streets to industrial pollution—are so extreme that Shell may be a model citizen setting higher standards for the country. "Shell is committed to the environment," she says, but "they also don't want the headline in the paper, the headline that says they're devastating the otherwise wild habitat surrounding their operations. It's true, we're embedded. But we think of that as an advantage."

The diminishing Gabon oil reserves are running out, stimulating local interest in the region's biodiversity. "The Gabonese are starting to look at their national parks and national resources as something that will be valuable," says Korte. "The Smithsonian can help with that cause." Meanwhile, Shell keeps poachers far from nesting sea turtles. "Even the sixty-kilometers-an-hour speed limit is good for wildlife," acknowledges Lisa Korte, "because driving at faster speeds you have a greater chance of hitting animals. There are some things that Shell does that actually work to the advantage of wildlife and biodiversity here."

Watching for elephants and keeping track of the speedometer—sixty clicks is just over thirty-five miles an hour—I drive across the Shell concession to the export terminal and office of Thoty Moussotsi Ikapi, the executive who heads Shell's Gabon environment section—a title that sounds to me like an oxymoron. The sign across from his building is an ominous reminder. EMERGENCY EVACUATION OF THE TERMINAL, screams its warning of action to take if the pipelines burst or the storage tanks explode in flames. "The general Terminal alarm is three long sirens. All staff within the Terminal boundary fence should secure their workplace and proceed to the muster point at the Terminal gate."

A biochemist by training, Ikapi engaged in HIV/AIDS research in South Africa before joining Shell in his native Gabon. A round-faced man with more hair on his chin than his head, Ikapi expresses a passion for the wildlife cohabitating with the pipelines crisscrossing their habitat. "We mitigate all our impact," he says about his oil company's policies. "We do not need to kill or destroy just for the sake of extracting oil." I tell him about watching the olive ridleys hatch on the beach. "Oh, I want to see that!" The oil exec's enthusiasm is spontaneous. He's never experienced the magic? "No! They called me one time. I was like, 'I'm coming! I'm coming!' But I had an important meeting so I went another day. I sat there, waiting for the turtles to come out, but they never came."

"I was out there this morning," I tell Ikapi, "and there were some children and their mothers—Shell worker children—and one woman said, 'It's really important to bring the children because they get this exposure to nature and maybe when they grow up they'll try to compensate for what their parents did to the environment.'"

Ikapi is rendered speechless. "Oh, that's"—he eventually laughs—"that's so kind." He laughs again before enthusing about how he combines his humanity with his employment. "Knowing that we're sponsoring the activities of those looking after the turtles, I feel just proud of being part of this company." Shell supports Ibonga, the NGO that operates the beach patrols. "I know that I bring an added value to the environment, to the universe."

"To those little turtles," I offer.

"Those little turtles, yeah," he agrees.

"Why are we attracted to and captivated by those turtles?"

"Their longevity," is his quick answer. "I think their longevity brings out humility in people." And that brings him back to his corporate parent. "We now have a role working in God's hands to continue this longevity of turtle." Or did. Not too long after my visit Shell lost interest and sold its onshore Gabon interests to the Carlyle Group, the massive and controversial U.S.-based private equity company.[4]

— *Silly Scientist* —

Back at the Smithsonian Gamba research station the humility theme continues at its intimate museum space. Cameroon-born lawyer Elie Tobi, a self-taught herpetologist, serves as assistant curator. While I looked at the specimens on display he told me that the customs of his tribe—the Bamiléké—forbid eating turtles. "It is believed they are spirit," he offers as a beginning explanation. But there's more to his turtle story. The Bamiléké do not eat turtles, nor do they keep turtles as pets or companions.

Nonetheless, scientist Tobi kept four turtles at the lab where he worked. He was going to be absent from the lab for too long to leave the turtles alone, so he decided to release them into the wild. He let the first one out of the bucket that he used to carry them into the Gamba swamp. "The turtle stood there looking at me without moving."

What to do? Tobi decided to walk off. After a few paces he looked back and the turtle still was looking at him. He decided to return to the turtle and explain what was happening. "I came back and I talked to the turtle. I say, 'You are free. You can go now. This is where we caught you before. I am pretty sure you didn't forget your way.'"

The turtle still did not move.

"I say, 'Thank you very much for the time you spent with us.'" Then he added an apology. "I say, 'Forgive me. I'm pretty sure I didn't treat you as I treat myself. I'm pretty sure that some days you slept without eating. You were hungry but you didn't eat. I'm really sorry for that. So, please, forgive me, but I'm wishing you good luck in your new free life.' And I shut up. The turtle turned around and went into the swamp."

The scenario, he says, repeated with all four turtles.

"I started thinking that turtles probably are very close to human."

Close to human? This from a lawyer and the curator of the Smithsonian's Gabon collection?

"They understood me."

Understood?

"Yes. As a scientist I fear saying silly things. This turtle story can seem silly. But I am still a human being and I experienced it. I think it was a spiritual connection. There is," says Tobi, "a strong relationship between animals and humans." Nonetheless, he finds it necessary to repeat his caveat. "This is a very good story for a grandmother," he says. "I feel no science in it, so I am reluctant to share it."

But share it he did, and as promised, Elie Tobi got back in

touch with me after he checked with contacts in his Cameroon native village—and not just any contacts. He consulted two sources he considered credible: a Bamiléké princess and a Bamiléké prince. I wanted to know the basis of the Bamiléké tribal rule against harassing, molesting, and killing turtles—and the origin of the custom that forbids eating turtle.

"Turtles," Tobi reported, "are sacred in our village because they are considered divine." But not just divine as in heavenly. When there is a dispute between villagers, turtles divine—with their actions—who is telling the truth. Tobi offered an example. "If someone is accused by another person, the turtle will tell if the accusation is true or not." A third party picks up a turtle chosen for the task and talks with it about the matter in question. The turtle judge is introduced to the belligerents. The turtle is then placed on the ground between the two folks at odds. "The turtle will turn around and then run between the feet of the guilty party and make a poo there and then hide its legs and head in its shell. That is how the turtle tells the truth. The turtle is used for justice."

Along with these evocative details, Elie Tobi sent a short video clip of such a truth-telling ceremony. In it a villager is holding a turtle that's about the size of a submarine sandwich. The woman with the turtle points at it repeatedly as she explains the case before it. She places it on the ground, continuing to point at it, poking it with her index finger. Villagers watch as the turtle sits motionless for a few seconds and then begins walking. The clip ends with the turtle still en route on its mission to serve justice.

• Happy Fred •

When I worry about Fred not eating and about his other bodily functions, I try to take comfort in Matt the turtle purveyor's coda to his list of turtle care instructions: "Fred is idiot-proof." This morning, when I came downstairs for my oatmeal and tea, Fred was in his house rustling around. I walked without making noise (I thought) over to his side of the living room and peered into his enclosure. There he was, in his big red water dish, bathing under his basking light, his head stretched out toward the warmth of the lamp, his wattled neck stretched almost smooth.

I pull a chair up to commune with Fred. He turns his head and looks at me with one big red eye, apparently decides I'm no threat and settles his head on the lip of his water dish, chin submerged. But when I lean toward him to take a closer look, his head stretches up again. There is no question in my mind that he detects my presence and my movements but doesn't perceive me as a threat. He's not pulling his head into his shell. He shows no signs of distress when I come to hang out with him. But he doesn't seem particularly interested either. That's okay with me. I'm just glad he's moving around, tussling with his worm dinners. In fact watching him frolic in the water—albeit at extreme slo-mo—I'm going to characterize Fred this morning as happy. And that makes me his happy roommate.

— *Reality Check* —

"Beach patrolling is all good fun. But the real problem for turtles is bycatch." I'm back in Libreville, talking with old African wildlife hand, British-born Lee White. We're sitting on the veranda at his home in the heart of the capital city, and his young daughter is splashing away the equatorial heat in the family swimming pool. Zoologist White has been tramping through Africa for over a quarter of a century studying its wildlife, and when we talk he's speaking as director of Gabon's national parks. Turtles, he tells me with his Monty Python–like sense of the absurd, make ideal poster children for conservation. "An elephant to a girl is a big scary thing. Elephants eat people's crops and they crush villagers or they bite them. There are probably more elephants in Europe than there are in Africa because every little kid's got a cuddly elephant in their bedroom. Here in Africa, because elephants are a scary thing, we can't use them as the suitable flagship. But turtles and whales are great, because everybody has this romantic view of them, and there's absolutely no negative connotation, especially the turtles." On the contrary, there's romance connected with turtles. "To meet a leatherback at night on the beach is really quite spectacular for kids."

Gabon was Dr. White's destination for his Ph.D. research because of the country's extraordinary biodiversity. He stayed on, working as a conservationist, and he helped convince President Omar Bongo to sequester almost twelve thousand square miles of the country from logging and other extraction industries and to create a legacy of thirteen national parks.

"The biggest threat to turtles," he says, "is bycatch and it's probably been killing a couple of thousand turtles every year for the last twenty years." Bycatch, along with plastic bags and rogue logs. Walking along the tide line on Gabon's otherwise pristine coastline, I was shocked to see the perpetual trail of debris: plastic bags, flip-flops, water bottles, and all kinds of assorted garbage. This is not necessarily Gabonese waste, but junk washing ashore from the rest of the throwaway world. Logs that wash up on the beach keep turtles from digging their nests above the tide line, so their eggs wash out to sea with high tides. "If we could get rid of the logs then we'd have a lot fewer nests drowned out. We don't have enough people on the beaches to dig up every single nest that's below the high-water mark." Those logs are residue of the timber industry, either inventory that got away or was set free on purpose by log poachers.

Turtles and tortoises rarely show up in the roadside Gabonese bush-meat markets. "You'll occasionally see a tortoise hanging by the side of the road by its foot, alive, and they're selling it. They'll sell it for a couple thousand francs to eat. Kinixys— hinge-back tortoises." When White answers my questions, his eyes bore at me and his ruddy face is devoid of ready reactions to my queries. When he looks at me, I feel as if we were out in the wild, not sipping tea in his city house—as if I were prey he is sizing up for study, or worse.

Before I finish my tea and leave, he tells me stories about Gabonese who believe humans can transform themselves into other animals. He talks about a princess who charged White with turning himself into an elephant and eating her crops, and about a government wildlife worker who expressed fear that a conservationist would turn himself into an elephant and abandon him if hippos attacked during a trek they were on together. Elephants, tigers, but I heard no Gabonese folk stories about people becoming turtles.

— *Codifying Turtle Folk Medicine* —

Housed in a Gabon public building badly in need of repair are the labs where healers use ancient Gabonese potions and techniques to treat the ill. At the Institut de Pharmacie et de Médecine Traditionnelle the practitioners try to reconcile their traditional remedies with modern medicine. That's where Paul Okouma conducts his research. He says he uses the rigors of scientific methods to test folk protocols, protocols he is convinced work. One of his raw materials of choice is turtle carapace.

I find him on a hot and sweaty day about to take the afternoon off to go to a wedding. But he graciously unlocks his lab door to give me a tour and explain his work. Liter-size plastic bottles filled with liquids of various colors cover the lab counters. Jars are scattered about, jars repurposed from their original uses for food or drink. Okouma takes an old pickle jar—or whatever was in it before it became a vessel for science—and shows off the brown powder inside. "Turtle," he announces. It's ground carapace and turtle tailbone, says Okouma, and I never would have guessed what it's used to treat: external hemorrhoids.

Practitioner Okouma tells me he only works with turtle bones and the carapace. First he burns the shell and then he grinds it into a powder. Next he mixes the powder with almond oil. The result is "applied to the infirmity." He shows off the nearly empty jar as proof of the treatment's popularity with his patients.

Turtle "thighbones" serve a different role in Okouma's practice. He festoons a cord with the bones and ties it around the waist of children suffering from malnutrition. The thighbone belt is also prescribed for patients who seek long life. He mixes ground carapace with a variety of plants and bark to create a remedy for stomach problems. Okouma is nonspecific regarding the species he uses for his cures. But he insists he limits his practice to land turtles. No sea turtles and no tortoises. Just land turtles, and, he says, since there are "tens of thousands" of them and only a few are taken for his studies, he is convinced Gabonese turtle populations are not jeopardized by his work.

Tribes in Gabon don't use the turtles for medicinal purposes, according to Okouma. His experiments with turtle parts are his own inventions. He knows of some tribes that use live turtles as guides. Hunters place a turtle in the middle of a path in the forest, and believers follow the direction the turtle crawls expecting that is where they will find game. Tribes take turtles that die of natural causes and use the carapaces as totems to help them communicate with their tribal ancestors. Overall, says Okouma, turtles are revered and believed to be very wise, acting as natural communication bridges between animals and humans.

— *Chichewa Variation on Aesop* —

Twenty-four hours after leaving the wilds of Gabon—from hippos and elephants on the beach to ground carapace hemorrhoid treatments—I'm back in Paris, where I hear another African turtle tale in UNESCO's high-rise Paris headquarters. There, looking out on the Eiffel Tower, Fackson Banda works in the organization's Communication and Information Sector, a job influenced by lessons taught to him in Chichewa by his grandmother on the shores of Lake Malawi in Zambia. The story is similar to Aesop's tale; it goes like this.

Turtle's apparent weaknesses, such as his slow motion, disguise Turtle's cleverness. This can be seen in a race between Turtle and Hare. Before the race Turtle hides his look-alike relatives along the route. The match begins and Hare races ahead, leaving Turtle so far behind that Hare no longer sees him on the path. Hare stops to relax, thinking to himself, "I left Turtle so far back on the trail I can take a rest." But then he hears a voice taunting him from ahead, saying, "I'm still here!" Hare, confused, races ahead, and the scenario repeats until one of Turtle's relatives crosses the finish line ahead of Hare, who is frustrated and confused.

"We were taught values through these stories," says Banda, who remembers the turtle story but not others his grandmother told. "I wish I were a repository of them. I've forgotten too many of them, and I didn't share them with my children." Modernity intrudes, he says: "Now TV, books, and films teach values." Instead of telling the old Chichewa tales to the next generation, he tries to teach his offspring values by helping them interpret mass media messages. "Maybe that's a tragedy," he concedes.

— *Paris Turtle-Free Zone* —

I grab a baguette at a stand under the elevated Métro tracks at Grenelle and take the line 8 train under the Seine to the Place de la Concorde. I walk along rue de Rivoli—the sparkling capital city of the old empire (at least in this showplace neighborhood), such a radical contrast to the shabby colonial remnant of Libreville. I wait in a short line at the Louvre with my appointment ticket. *"Excusez-moi,"* says a soldier, his jaunty French beret offering a softening counterpoint to his camouflage uniform and assault rifle. He and other troopers walk past the line and on into the museum, their overt show of force in the Louvre the newly militarized look of the City of Light on this day just a couple of months after the random attacks on civilians that bloodied Paris in November 2015.

Once inside the museum I head directly to the sculpture galleries where Pierre Hébert's *Child Playing with a Turtle* resides. Photographs I've seen of the piece show a naked young boy perched atop a sea turtle, a makeshift bridle deep in the not-so-happy-looking turtle's jaws. The awkward pose suggests that the boy is about to call out a "Whoa" as he urges his white marble mount to a stop. Or maybe it's "Giddy-up" he's calling out and he's trying to get the turtle out of the gate. Hard to determine from pictures of the 1849 carving whether the two are coming or going. But once I finally locate the gallery where the actual statue lives, the door is blocked—the space is closed for renovation. I take it as a sign. Here I am in Paris! Here I am in the Louvre! And I'm fixated on finding an obscure sculpture by an obscure sculptor. I take the TRAVAUX EN COURS notice on the door as a sign directed at me personally. Time to take a break from my everything-turtle work in progress, a break to keep my penchant from becoming obsessive.

• Up Against the Wall, Fred •

Fred is climbing the walls. Again. He's scratching at the walls of his house much like the Russian tortoise at Petco. But he does not seem discontent. When I come over to check on him he looks up at me, stops climbing the walls, and hauls himself toward me high on his scaly legs, with his neck outstretched, his eyes checking me out. He's one active turtle this morning, strolling around his house, basking under his lamp, wandering into his dark room. Or is he restless?

I cannot imagine he knows I've been thinking about saying goodbye and about his upcoming return trip to Arizona.

Now he is up high on his hind legs, scratching with his front legs at the wall, reaching his head toward his basking lamp. It is disconcerting for me to watch after seeing him so passive for so long. But I'm hoping he's just enjoying a stretch, getting some exercise, seeking more warmth from the lamp. Now he's on his hot pad again, just basking. That makes me feel better about his state of mind—whatever mind he may enjoy.

NINE

THE PITIFUL CASUALTIES

— Hare-Brained Cowboy Meets Tortoise —

"If you ain't a Nevadan you ain't shit," reads the text on a chipped enamel lapel pin I wore with pride when I lived in the Battle Born state. Nevadans idolize the myth of the independent Western character, a character free of the burdens and restrictions so many governments place on their citizens. When I moved up to Nevada from California, I embraced that legend—from the gun rack in the cab of my '56 Chevy pickup to the go-cups (legal in those Wild West days) where the End of the Trail saloon bartenders in Dayton would pour my tequila-and-tonics to help fuel my after-the-parties drives home.

The official Nevada reptile is a crusty and wizened curmudgeon that looks somewhat like a sun-weathered and wrinkled Nevada cowboy. That state reptile, the Mojave desert tortoise (*Gopherus agassizii*), looks a lot like old Cliven Bundy, the addled cattle rancher who likes to sport a copy of the U.S. Constitution in his cowboy shirt pocket despite the fact that, based on his seemingly endless orations to TV cameras in the spring 2014, he couldn't pass a grammar school civics course.

For those lucky enough to have missed Bundy's fifteen minutes of Warholian fame (which unfortunately lasted much more than his allotted fifteen minutes), here's a brief primer on his antics (and a tip of the sweat-stained Stetson to reporter Jaime Fuller at *The Washington Post*, who assembled a Bundy timeline).[1] The drama dates to 1989 when the desert tortoise was added to the Endangered Species List. Four years later the Bureau of Land Management set aside tracts of federal land—including one known as Gold Butte in southern Nevada—in an effort to preserve habitat for the tortoise. Ranchers—including Bundy—were ordered to end cattle grazing on those public lands—grazing that required both a permit and a fee. Bundy continued to run his cattle on the protected Gold Butte public land in defiance of the rules. He ceased paying federal grazing fees and he refused to pay the fines levied against him for the violations. Bundy kept a-grazin'—claiming rights to the land based on his family's long history using it, along with his rejection of federal government authority over public land. In the summer of 2013 the BLM informed him that its officers were going to round up his livestock.

That's the point when Bundy declared his "range war" against federal authorities. "It's not about the tortoise!" brayed camouflage-clad riflemen who made the pilgrimage to the Bundy ranch to join his posse. But it clearly *was* about the tortoise as the threat of a gunfight between the feds and the outlaws escalated. Meanwhile Rob Mrowka, a staff ecologist at the Center for Biological Diversity, explained why his organization threatened to sue the government if the BLM continued to allow Bundy's cattle to trample the tortoise preserve. "The Gold Butte area has been officially designated as critical habitat for threatened tortoises," he and several colleagues wrote in a public missive, "meaning it is essential to their long-term survival as a species."

A cardboard sign appeared by the side of a Nevada desert road, the scrawls in black marking pen nixing the BLM with the international diagonal line for no. "We the people this is our land!" yelled the anonymous sign painter, adding the silly footnote, "Take your tortoises and go home." Not noted in the sign is the fact that the tortoises predate the Bundys as Nevada landholders by at least a few millennia and the local tribes, of course, lived on the land long before the first white settlers came west. The Bundy family ranchland was part of Mexico until President Polk instigated a war and seized it for the United States in the name of Manifest Destiny. The Treaty of Guadalupe Hidalgo ceded the Gold Butte territory to the U.S. government in 1848. Bundy dated his family's history on the land in question only back to 1877—that's yesterday by tribal, Mexican, and tortoise standards.

The desert tortoise, drawled Bundy on the Fox News *Hannity* show, is an animal used by the federal bureaucracy to exert control over public land. "The desert tortoise does use the same habitat and the same type of feed as livestock," he acknowledged, "so there could be encounters if my cattle ate all the feed or the tortoise ate all the feed. But that's not the case here on this desert. It's impossible to use up the feed for the desert tortoise," he asserted to the TV audience, with his Western credibility enhanced by the wide vista of sage, sand, and distant mountains in the picture. "I'll never get it," he confessed to the *Las Vegas Review-Journal*. "If it weren't for our cattle, there'd be more brush fires out there. The tortoises eat the cow manure, too. It's filled with protein."

Bullshit (so to speak), is the retort from conservationists working to protect the desert tortoise. "That's just nonsense!" Enter Dr. Michael Connor, biologist, California director of the Western Watersheds Project, and, previously, executive director of the Desert Tortoise Preserve Committee, an organization dedicated to promoting *Gopherus agassizii* in a wild and native status. He laughed about the claim that tortoises thrive on cattle dung. Connor's expertise includes public lands management and habitat conservation. He lobbied the federal government to list the desert tortoise as an endangered species, and in 1990 it was officially listed as threatened. He was incredulous when I told him the dung-as-tortoise-food theory. "Why doesn't Mr. Bundy eat it?" he suggested. "That's just rancher crap. It's just pure nonsense."

Gold Butte is a tortoise habitat that Connor has trekked and knows well. No matter the territory, he's convinced "cattle grazing is counterproductive for tortoise everywhere. But it's especially true there, where the number of cows that are being turned out is something like three or four times higher than even the BLM would allow." There are so many cattle grazing at Gold Butte, according to his survey, "that there's basically nothing left for tortoises to eat." He termed the landscape severely overgrazed. "It's not just a question of a few cows wandering around. We're talking about an area that can barely support any cows at all." Besides cattle eating vegetation sought after by tortoises, Connor observed what he called "major hoof trampling, a lot of soil compaction." Get rid of all the cows grazing on public land, is Connor's advice, and the price of steak at the supermarket will not change because of it. "Public lands ranching accounts for a tiny percentage of the total cattle in this country. Unfortunately, what it does do is account for an enormous percentage of the damage done to public lands, particularly in the West."

The 2011 U.S. Fish and Wildlife Service recovery plan for the Mojave desert tortoise corroborates many of Michael Connor's arguments, albeit with less colorful language. "Grazing by livestock (cattle and sheep) affects desert tortoises through crushing animals or their burrows, destroying or altering vegetation (which may introduce weeds and change the fire regime), altering soil, and competing for food," concluded the report. "The Mojave Desert is neither highly productive, nor is it an environment which historically supported native cattle-sized grazers." The report called for minimizing the impact to tortoises within conservation areas such as Gold Butte, and it specifically called for the removal of trespassing cattle like those of Cliven Bundy.

"I don't think we should allow anything to go extinct," says

Michael Connor. "Why would we want the desert tortoise to go extinct? It's a keystone species in the desert. It's a major indicator of desert health. Desert tortoises in that area," he says about Gold Butte, "used to be much more abundant. Desert tortoises live in the ground. They dig burrows in the ground. They have an extensive burrow system. An individual desert tortoise has multiple burrows. Other species of animal use those burrows. It has an impact on local hydrology. Desert tortoises are little ecosystem engineers out there. Get rid of those and who knows what's going to happen to everything else."

Before we said our good-byes, Michael Connor lectured me about the desert tortoise. "It's a very charismatic animal," he said. "Have you ever seen desert tortoises out in the wild and watched them behave? They're quite remarkable. You're in the middle of an area where there is very little vegetation most of the year, so it can be pretty dry. There is wind blowing. And you see these tortoises moving around and running around." Running around? I always thought that was the hare. Hard to imagine lumbering tortoises running around. "They clearly know every nook and cranny of the area they live in. It's quite remarkable. They have an amazing knowledge of their habitat. They interact with people. Not always a good thing, but they do. Some of them, when they see you coming, will actually come over and step in your shade, as though you are a mobile Joshua tree." Nice image. "Others will run away." Connor is an ambassador. "Everybody likes tortoises," he offered with enthusiasm. "The majority of people consider them to be inoffensive." That's a left-handed compliment. Yet he said again, "They're charismatic."

That's more than can be said about Cliven Bundy. As he ranted and raved about his twisted interpretation of the American law and the U.S. Constitution, the band of misfits he attracted spoiled for a fight with the BLM officials assigned the nasty task of rounding up Bundy's cattle. To defuse the crisis, the cattle that the BLM wranglers had rounded up were turned loose; the feds backed off from Bunkerville to strategize further. Some Bundyites jumped off his bandwagon after *New York Times* reporter Adam Nagourney reported details from a news conference Bundy presided over (with Nagourney and a photographer the only attendees).[2]

Pointing to the departing feds, Bundy declared victory and announced to Nagourney in an unexpected non sequitur, "I want to tell you one more thing I know about the Negro." He wandered down memory lane, recounting a public housing project in Las Vegas he saw in passing. "In front of that government house," pontificated the daft cowboy, "the door was usually open and the older people and the kids—and there is always at least a half a dozen people sitting on the porch—they didn't have nothing to do. They didn't have nothing for their kids to do. They didn't have nothing for their young girls to do. And because they were basically on government subsidy, so now what do they do? They abort their young children, they put their young men in jail, because they never learned to pick cotton. I've often wondered, are they better off as slaves, picking cotton and having a family life and doing things, or are they better off under government subsidy? They didn't get no more freedom. They got less freedom." It was a speech that eroded Cliven Bundy's folk hero status. Nonetheless, four years later a federal judge—citing errors by prosecutors—tossed out the charges Bundy faced: conspiracy, assault, threats against the government, firearms violations, and obstruction of justice. He walked out of a Las Vegas courthouse a free man, his cattle still competing with desert tortoises on the public lands of Gold Butte. From my office at the University of Oregon I contemplate the words on my old lapel pin—"If you ain't a Nevadan, you ain't shit." As a displaced Nevadan observing the Bunkerville scene from my academic ivory tower, I'm convinced that Cliven Bundy, despite his alleged Nevada pedigree going back to 1877, doesn't exhibit the open-minded spirit of Nevada—hence, he ain't a Nevadan. And, according to my lapel pin, we all know what that means.

— *Straying Sea Turtles* —

Climate change is a threat to sea turtles. Fishing in the Pacific off the San Francisco coast in the late summer of 2014, Roger Thomas, the captain of the *Salty Lady* party boat, was surprised when he hooked a sea turtle far from its usual haunts. "We were trolling for salmon," he recounted to me when we talked after the catch. "The starboard bow rod hooked something. We didn't know what it was because it was underwater." The line that hooked the something was about thirty feet into the water. The deckhand brought the line to one of the tourists on board "so the customer could get nice play out of it and bring it up to the boat." The creature pulled back. "Immediately it became very obvious that this was something much heavier than a salmon. The deckhand had to help the lady." When the turtle surfaced, Captain Thomas' boatload of tourists began screaming with excitement and concern for the health of the turtle. It was a Pacific green turtle (*Chelonia mydas*), endangered and far from its usual range south of the U.S.-Mexican border.

The crew netted the turtle, brought it aboard the *Salty Lady*, and removed the fishing hook. It looked unhurt, so they dumped it back into the Pacific. "It was very active and it dove. We saw it come up at a distance. It was fine." In his thirty-eight years of plying the waters off California, this was the first Pacific green spotted by Captain Thomas. "Kids grow up with little turtles," he said, trying to explain to me the thrill on board his boat that day, "and when you see a turtle in the ocean, it's a monster in comparison with that little turtle they were raised with. Maybe that's why people feel so close to sea turtles, because they've seen little turtles on land."

Several hundred miles farther north, I visit two other stray sea turtles—olive ridleys—at the Oregon Coast Aquarium in New-port. Both were found beached following severe late 2015 storms, malnourished, comatose, and suffering from hypothermia. Lightning and Thunder, the staff named them, not because they're adopting them as pets or for exhibits at the aquarium but to distinguish them for the logs documenting attempts at rehabilitating the two. Aquarist Mark Murray is nursing them, trying to help them regain their health and strength. "It's not just like a fish," he says as we look in on his patients, segregated in their own tanks. "Not many people are emotionally connected to most fish, but then when they see a sea turtle most people get really excited." Not Murray, who worked for years with sea turtles along the Carolina coast, where they are a relatively com-mon sight. His satisfaction comes from rehabilitating those in need. "To get two more turtles back out into the wild to breed, whatever we can do to help, is the thing I like the most." He figures these two, foundering so far north, were chasing their next meals—anchovies and sardines—in unusually warm Pacific waters off Oregon. "Younger turtles, who don't know not to go this far north, will follow the food. When winter comes and a storm hits, they can't get south quick enough and they get cold-stunned." They try to swim south to warmer water, Murray says, but they burn up their fat reserves and, exhausted, they can't even digest food they've already eaten, let alone forage for more. No longer able to swim, they wash up on the beach and languish unless rescued.

Therapy for the beached turtles starts with getting them back into water. Murray and his crew assess what ails them and what treatment they need, and they prompt the turtles to eat again with shrimp and squid for protein along with bony fish for the calcium they need for shell growth.

"Not really," is Murray's response when I ask if he's bonding with the two rescues. "I'm glad that they're both swimming and breathing, that they're alive. But it's definitely not like a dog, it's not a pet. It's not like when I come home and I see my dog and my dog is super excited to see me. You know the endgame is that they're going to be released. That's where the emotion is, to see them released." We're watching Lightning looking up at us, and Murray says it's not because she wants to cuddle. The feeling is mutual. "I don't have a desire to get in the water and swim with her." Lightning isn't looking for a swimming partner either. But she knows when Murray shows up at the tank it's feeding time. "They're messy eaters," he notes as they mash the cut-up herring he tosses toward their waiting beaks.

"You're not tempted to reach down and scratch her head and inquire, 'How are you, sweetheart?'"

"No," Murray tells me, "because there's no reaction from a turtle to that. They don't like petting or anything like that. I've been to a lot of sea turtle releases at the beach. You put them on the sand and they don't look back."

They're loners. They don't hang out with their mates. They abandon their offspring before they hatch. Why should they care about Mark Murray even if he's trying to save their lives? After a few months in his care, the Coast Guard took advantage of a routine training mission to airlift Lightning and Thunder south to SeaWorld in San Diego for further rehabilitation in a warmer clime. Lightning continued to recover, but a month after arriving at SeaWorld, Thunder succumbed to the traumas she suffered. "The odds of saving stranded animals are low," acknowledged the Oregon Coast Aquarium's director of animal husbandry, Jim Burke. Sad but realistic, Burke said after Thunder died, "Our team will continue efforts to save injured, threatened, and endangered species for the chance of boosting their low populations."[3]

• Fred as Friend •

"Hello, Fred." Is he looking at me with happiness for some fleeting companionship and attention before I head off to the kitchen to make breakfast for myself? Happiness is hard enough to define for myself, I think, as I look back over at Fred. "Hi, Fred," I say again, and he turns his head in my direction with no hurry and stretches his neck out—toward me or the warmth of the heat lamp? Impossible to know.

Now he's ever so slowly turning directly toward me, his paws (or claws) scraping against his hot pad as he pivots an inch or so and stretches his neck out toward me. He's definitely looking up at me, not the lamp. It's been a couple of days since he ate. Perhaps he does recognize me, at least as a vending machine for meals. Or he's just hopeful that this other being in his midst might translate somehow into food. Whatever—it works. I'm off to the kitchen to get grub for Fred before I make my oatmeal.

— *Zoo Beat* —

Just before I fly into Buenos Aires on turtle business, the city zoo closes. "Today this place generates more sadness than happiness," Mayor Horacio Rodríguez Larreta said when he announced that the zoo would be replaced by a park populated only by animals that choose to wander its gardens. "Captivity is degrading for the animals," proclaimed the mayor.[4] When I stop by the front gate a few days later, the stone-faced lion still stares out from his regal perch at the top of the entrance arch under the welcoming words *Jardín Zoológico*. A stone duck is headed on the arch toward the lion, its stone wings suggesting flight, while a live duck stares at me from behind the closed gate, closed with a fat chain locked around it. At the shuttered ticket office I see a sign sponsored by Coca-Cola that reads: SALIDA. HASTA PRONTO! "Exit" is correct, but "See you soon"? Probably not. According to Mayor Rodríguez Larreta, all the animals healthy enough to leave are headed for sanctuaries; for those animals deemed able to survive without help, eventual freedom in the wild awaits. Presumably that includes the turtles and tortoises, but not the jungle tortoise made famous by the Argentine author Horacio Quiroga in his 1918 story for children, *The Giant Tortoise's Golden Rule*. It is a sweet interspecies tale. A hunter collecting specimens in the jungle for the Buenos Aires Zoo saves a tortoise mauled by a panther. The hunter nurses the severely injured tortoise back to health, but then he himself gets deathly ill. The grateful tortoise packs the sick hunter on her shell and manages to walk to Buenos Aires to find a doctor, who cures the hunter. As its reward, the tortoise is provided a happy home in the zoo, "with a tank of water in the front yard, where she could swim if she wanted to. She was allowed to wander at will over all the gardens of the Zoo," Quiroga writes, "though she spent a large part of her time near the monkey house, where there was the most to eat. And she is still living there."[5] Maybe. The tortoise in the story could have been alive in 2016 when the zoo closed, a spry survivor, well over ninety-eight years old.

"The most important thing is breaking with the model of captivity and exhibition," animal rights lawyer Gerardo Biglia told reporters when the mayor announced the zoo was closing after 140 years of entertaining visitors. "I think there is change coming," he predicted, "because kids nowadays consider it obvious that it's wrong for animals to be caged." Sheets of rain wash Córdoba Street a few days later as I head for Biglia's downtown Buenos Aires office. He welcomes me into a small conference room, the walls lined with law books. "We've arrived late to help many of the animals," he laments, but expresses hope that, as he puts it, society will move from "talk about human rights to fundamental rights, with everybody included." When he says "everybody," he means animals. Biglia helped argue the successful Argentine court case (a global first) that resulted in zoo resident Sandra, an orangutan, being ruled a "nonhuman person" and hence with rights to her freedom (she is one of the animals deemed too infirm to move and will live out her life at the zoo, though not as an exhibit for the "human persons" to gawk at).

Biglia exudes seriousness as he philosophizes about animal rights, his salt-and-pepper beard and over-the-ears hair adding a professorial hue to his impassioned speech. Of course, the court ruling that orangutans are persons was a decision not universally embraced, and Biglia agrees that some sort of Aristotle-like *scala naturae* must be accepted as reality. "My five-year-old daughter cannot sign a contract. All species have rights that should be guaranteed," he insists, but their needs and desire differ. "Cows do not have an interest in the right to an education or to take dance lessons. But they do have the minimum right not to be treated as objects." His dancing cows—what an image!—example leads me to ask where he figures turtles and tortoises fit into the rights debate. "One thing they all share," Biglia says about turtles, tortoises, and the entire animal kingdom, "one thing they would claim from us, is the right to their habitat—for us to stop destroying their ecosystems. In South America we are destroying forests to create a monoculture. We are destroying many species." His words makes me think of the graffiti I've seen on the Argentine cityscape: MONSANTO = GENOCIDO.

In Argentina, *asado* is the word for "barbecue": both the way the famous Argentine beef is cooked and the party that comes with it. But the land for grazing cattle is disappearing, overtaken by soy—20 million hectares in 2015,[6] with massive increases planned and expected.[7] And soy means Monsanto, with its patented Roundup Ready seed that sprouts bean plants immune to its glyphosate-laced herbicide Roundup. Hence the graffiti. This march of soy (and corn) into Argentina's grazing and wild lands is encroaching rapidly on the habitat of *Chelonoidis chilensis,* the Chaco tortoise.

— *Roadside Turtle Merchants* —

Days later I'm speeding through some of those soy fields in a rattling rented Toyota, heading north toward the Chaco, the arid Argentine region bordering Paraguay and Bolivia. The last

time I was in the Chaco I was on the Bolivian side, chasing organic beans, not the eponymous Chaco tortoise.

The truck traffic on *Ruta* 34 is nasty, one long tractor-trailer
after another hauling south fast on the two-lane blacktop, passing just inches from my Toyota, the tiny car buffeted in their
wake like a nervous pony. This is gaucho land. Twelve hours on
the road, with the speedometer often hitting a shaky 140 clicks
per hour, across the pampas, Argentina's big-sky country, its
breadbasket, home to its famous beef, and—the farther north I
go—lonesome territory. When the gas gauge dips below half,
I fill up the Toyota. As I cross into Santiago del Estero Province
trees disappear for several miles at a stretch, leaving just scrub
nothingness sweeping toward the faraway horizon. But much of
the landscape looks comforting and familiar, like the deserts of
the North American West.

Just north of the village of Icaño I spot the first sign of a Chaco
tortoise. It's a hand-painted sign on a scrap of wood, the white
block letters calling out: SE VENDE TORTUGAS. I pull onto the
gravel shoulder, and the close-passing trucks make my car shud-
der. Beyond the sign animal pelts hang on display; on a makeshift table is a row of roosters hewn from what look like pine
logs. I wait for a break in the truck traffic, get out of the car,
and walk toward a hut. As I get closer, a woman appears and
comes out of the shack. She's wearing faded blue jeans, with
a white scarf and a black windbreaker to fend off the winter
wind. When I ask her about *tortugas* she produces a Chaco
tortoise in her palm. It's a cute little fellow and looks healthy
enough, moving about on her hand despite what probably
should be its hibernation time.

I admire it and ask her name and the price.

"Three hundred pesos," Lisa says, and tells me she already
sold three this day.

"Who buys them?" I ask.

"Tourists. The locals are used to them."

Ruta 34 is the main highway connecting Buenos Aires and the Andean colonial getaway city of Salta. She tells me the tourists buy them to keep as pets. I query her about how long they live when they're uprooted from their home turf and taken to the likes of a Buenos Aires apartment.

"I don't know," Lisa says, and then speculates with enthusiasm, "a hundred years!"

"Really?"

"Yes! My *abuela* had one and it was still alive years after she died."

"How old is this little guy?"

"A year or two," she says. "They get big." With her hands she indicates the size of a Frisbee.

Lisa doesn't seem to mind my questions or my camera. She says she does not catch the tortoises herself but buys them from collectors who find them to the east of Icaño. There is nothing clandestine about Lisa's illegal roadside pet store. She shows me an old tire filled with sand and says that during warmer times of year it's filled with the little Chacos. A handful of little kids hover around asking for *caramelos*. The scene is mellow. Lisa is not trying to convince me to buy a tortoise (or a puma skin or a wooden rooster); she's just offering them when I ask. And the *niños* are not insistent or demanding like Chiclets peddlers in Tijuana; they're just asking for the candies.

Back in the Toyota I continue north past a parade of stands that look just like Lisa's: the hand-painted sign offering *tortugas* and rustic wood carvings. A few kilometers up the road is the next gas stop, Colonia Dora. I was ready to rough it in a hotel as rustic as Lisa's wooden roosters, but like a mirage the Samay Hotel appears at the intersection of Routes 34 and 92. It's a self-described boutique hotel, the bed luxurious, the bathtub huge, and the internet connection slow but steady. It's just the oasis I need. After sleeping off the exhausting drive, at dawn I head for the lobby serenaded by a chorus of songbirds.

The night desk clerk, born and raised in Colonia Dora and happy for some company, proves a talkative and credible informant. Most of the indigenous people in Santiago del Estero are poor, he reports, with few opportunities to earn a living. They set up the stands I saw along Route 34 and—especially during the winter school holidays, when tourist traffic is heavy—they hope to flag down travelers and sell *tortugas* to them for 100 to 200 pesos. (Lisa wanted 300!) Entrepreneurs favor the particular stretch of highway between Icaño and Colonia Dora because traffic slows when it hits the villages. "It's not legal and it's not illegal," is how the desk clerk pragmatically describes the trade. The local police know the peddlers must somehow try to earn a living, and so they look past the SE VENDE TORTUGAS signs.

After breakfast I head back to comparison shop at other stands near Lisa's. Just south of Colonia Dora a woman bundled up against the cold meets me on the path leading from her hut. I'm curious about her sign selling *arrope*; she says *arrope* is the Quechua word for fruit made into marmalade. I ask her the Quechua word for *tortuga*. *Hguolu*, she says. She writes it in my notebook and offers to call a woman with *hguolu* for sale. Out comes the mobile phone; she speaks a few words of Quechua into it and suggests I wait. While I do she offers to teach me more Quechua words. She writes the words she chooses as she says them, adding the Spanish translation for my vocabulary lesson: handsome man and beautiful woman.

Shuffling along another path appears a chunky teenage girl accompanied by two exuberant bouncing puppies. As I say goodbye to my Quechua instructor, she offers me her phone number in case I want to learn more words. The teenager is carrying a tattered cardboard carton that's seen better days. Inside are six little Chaco tortoises, one trying, or so it seems, to find shelter from the cold under a blanket.

"What type of *tortuga* are these?" I ask.

"*Tortugas de tierra*," is the answer—land tortoises. It's the same response I got from Lisa. What's the scientific name, I ask the girl. "Land tortoise," she repeats. The Chacos in the box look lethargic; perhaps the girl has just disturbed their hibernation. She wants 400 pesos for each. "Up on the other side of the road," she points toward Lisa's stand, "they'll be 600." She wants only 200 pesos for bigger ones, saying that as they grow bigger they're worth less. Why? It's a price structure she's at a loss to explain. But at the next stand where I stop, a boy—clad in surfer shorts with a flower motif, his hands stuffed in his sweatshirt pockets— tells me the smaller Chacos cost more because they're harder to find.

I return to Lisa's stand with a pocket full of *caramelos*, buy a wooden rooster from her uncle, and hand out candies to the kiddies, who run off smiling to play under the hanging pelts.

Selling Chaco tortoises is illegal. The Chaco is identified as "vulnerable" on the International Union for Conservation of Nature's Red List because of both habitat loss and a thriving pet trade.[8]

The frustrations of law enforcement are fueled by the crises facing victimized turtles.

• Talking to Fred, Again •

"Hi, Fred." I am not quite sure what I expect when I hail Fred in the morning. He does look up at me when I pull up a chair to greet him and contemplate him—and hence life. He is more and more mesmerizing. I can look at him and find myself enthralled, although I'm not sure with what. The unknown (and unknowable)? Perhaps it is his subtle act. He blinks. He moves his head toward me and extends it when I move my head. He usually wins our stare-down contests. If patience is a virtue, Fred is one virtuous turtle. "Just give me my heating pad, basking lamp, and a night crawler," he might be thinking. "I'm happy to sit under this lamp looking at you for as long as you look at me." Or maybe, "I hope that big guy staring at me doesn't stuff me in a sock, pack me into a box, and call ShipYourReptiles.com for an airplane ride back to Arizona." Or maybe he'd like a trip back to the desert.

Fred sits with his right front leg extended, his five claws spread out as if he's ready for a manicure at a nail salon. His left leg is tucked back against his side with his palm up. He looks like he's in midstroke swimming the crawl across the heating pad except that his back legs don't look ready to kick; they're all but hidden under his shell. Just as I write this, out kicks his left leg. I refuse to believe Fred made that move in response to what I'm writing and thinking. And yet it was one of those eerie coincidences that help explain how easy it could be to believe there's some connection between us.

TEN

THE CONFLICTED PUBLIC AND THE DEDICATED CONSERVATIONISTS

— Turtle Abuse —

The foul-smelling apartment is in a middle-class neighborhood just a few blocks from the Mexico City airport. The owner's sister pleads with the Humane Society to break down the door. Instead, with police protection, veterinarians using keys obtained from the family's lawyer, open the door and try to stifle their nausea. The anonymous tipster was correct. Putrid Tupperware tubs piled on top of each other cover the kitchen floor and counters, tubs crowded with turtles languishing in dirty water. Some are obviously injured or sick; others appear healthy. When offered food, all prove ravenously hungry.

The doctors work fast. "It was shocking and creepy," veterinarian Claudia Edwards, the executive director of the Humane Society International/Mexico, tells me later. "The apartment was blacked out with curtains." Portraits of Che and Fidel adorn the walls, along with snapshots of the former occupant posing in Cuba wearing a Che T-shirt. "We thought the softshell was dead. It was upside down in dirty water." Squawking and singing at the overwhelmed vets are almost a hundred birds, some in cages and some not—canaries and parrots. "It was very, very sad." Out into a waiting van go the turtles. The vets want the animals treated at a clinic as soon as possible, but they're also worried that the turtle abuser may interrupt their rescue. Mexico's notoriously violent drug and gun cartels infiltrate the global illegal animal trade—and turtles, because of their specialized characteristics, translate into easy money—docile to transport and prized by collectors. The getaway is a Keystone Kops fiasco. Unfamiliar with the borrowed truck, the rescuers fill its tank with gas. The van runs on diesel and it spits to a stop. This day, Mexico's Independence Day, they are in luck. No one is following. The twenty turtles are saved without confrontation. Only three of the birds are dead.

A few days later, the turtles are stabilized at the Zacango Zoo in Toluca and Dr. Edwards is getting to know them. "Every turtle is unique," she tells me as we drive out of Mexico City's smoggy traffic jams and into the rural landscape around the zoo. "They have different likes and dislikes and abilities." I had my doubts. They still looked like nothing more than shells with legs and heads to me when she and I talked. I'd yet to meet Fred. "It's like a fingerprint in their brain and their personality," says Dr. Edwards. "Even animals in the same litter are different."

I appreciate that my old dog, Amigo, was unique. Even if he wouldn't fetch my slippers or bring in the newspaper, he was like no other. The very fact that he did whatever he chose to do is indicative of his individuality—from falling asleep outside in Nevada storms until he was completely covered with snow to running off across the desert chasing rabbits in those pre-leash-law days. But Amigo was a dog, not a slow-walking, cold-blooded shell.

"They have different likes and dislikes," insists Dr. Edwards about the turtles in her care. "These differences are the basis of evolution, so it makes sense." She describes one of the turtles as shy and quiet, a turtle that allowed her to pet it. "Another, a softshell, she was grumpy." A grumpy turtle? Dr. Edwards insists it's true. "She does what she wants to do. If I grab her and put her in the water, she splashes me with her back legs, like she is very angry, like she is saying, 'Leave me alone, please!' She speaks with her body language." Coincidence, I suggest? Not a chance, says Dr. Edwards. "The angry one is always angry; the shy one always appreciates a caress."

The sick among the rescued turtles are being treated for eye damage, for fungus growths, for herpes infections. Treatment includes plenty of food, a clean environment, and stress-free time in the warm Mexican sunshine. When the turtles were sunning themselves, Dr. Edwards perceived happiness—turtle happiness. They faced the sun. "I think they enjoyed the freedom, the air, the sun." During the night the rescued turtles were kept in tubs without water to keep their body temperatures from getting too cold. When the turtles were first in the tubs, they tried to escape. "They were not happy." But when the tubs were filled with warm water and placed out in the fresh daytime air, "they relaxed. They didn't try to get out of the tubs. They enjoyed themselves."

— *Personality or Biology?* —

Maybe the turtles simply respond physiologically to their basic needs being fulfilled, and it only appears to us humancentric observers that they experience stuff that we experience—like happiness. Nope, says Dr. Edwards. "I don't think the animals do anything they don't want to do. They are not machines. That's why I hate the word 'instinct.' I think animals have different options." It's not that the turtles are programmed like computers, she says, and that when the sun is out they must bask in it. "Animals choose if they want to do A or B or C or D." Dogs don't always chase cats, is her immediate example. "They look for options. They enjoy different things, and they show them to you. Some like to be with you, some don't. It's their choice."

We're talking, stuck in the slow traffic, and I imagine us being observed and studied by some other creature looking down at this mass of cars filled with creatures like me, creatures confined in metal boxes moving alongside other creatures confined in other metal boxes, all heading toward an urban mess of buildings jammed with more creatures like us. We're not stopping for a picnic in the lush and lovely countryside we're bypassing. It might make that creature watching me and my fellows wonder if the ability to think for ourselves is an option we can employ. Do we know how to find happiness if we choose a traffic jam over an idyll in the bucolic countryside? "But somebody is stopping for a picnic," Dr. Edwards points out, and she's convinced the same is true for the turtles.

Bioethics is Claudia Edwards' teaching focus at Mexico City's National Autonomous University of Mexico. She is a cat specialist, but she's dedicated to working with whatever animals need her care. "My own mother asked me, 'Why help animals in Mexico where so many people are in need?' I help people through animals," is her ready explanation. "If children are raised with animals, they learn empathy, compassion, duty, and responsibility." Society benefits, says the doctor. "People who are violent with animals, they are violent with people." Yet her mother, who, she says, delights in a fancy lifestyle, couldn't believe Dr. Edwards came home from her veterinary studies perfumed with animal smells. No matter. "We humans, we have voices. We have arms and we have legs. We can say, 'This hurts me,' or 'I am hungry. I don't have anything to eat.' But a turtle in a house in a dirty Tupperware tub, how does he say, 'I feel sad,' or 'I feel hungry'? A kid maybe can groan and cry and somebody can help. But a turtle?"

The fog has lifted, but it's still a cool morning when we arrive at the Zacango zoo. We walk quickly past its mission-style façade and into the sprawling verdant grounds. We pass the elephant and the camels, listen to orchestral birdsong, and glance at a massive capybara, weighing maybe 150 pounds—one big rat cousin. We arrive at the *herpetario* and meet its director, veterinarian Augustín Álvarez. Dr. Álvarez and his crew are busy trying to nurse the rescued twenty turtles back to health.

"We work in the worst area of the zoo," he tells me. Nevertheless, he strikes me as a jolly guy as soon as we meet. He smiles, he laughs, and he makes light of battles he faces educating the public about the value of turtles and the other animals living in his reptile house. Why is the reptile house the worst area in the zoo? He immediately answers his own question. "When someone says a panda is on the verge of extinction, everyone wants to help and contribute because they are fat and fuzzy and they look like stuffed animals." Not so with slithering, creeping, scaly reptiles: "People generally don't want to hear anything about them, even if they're on the verge of becoming extinct." He's convinced that most people think it's better if these animals cease to exist because "why would we want to conserve a venomous creature like a rattlesnake?" That attitude, believes Dr. Álvarez, extends across the entire class of reptiles, turtles and tortoises included. Hence his reptile house slogan: "Let's turn fear into fascination."

Turtles are "mysterious, cold animals," says Dr. Álvarez, who is himself anything but. His cherubic face exudes a warm openness that is the antithesis of the creepy reptiles in his charge, and his readiness to explain and philosophize is the opposite of mysterious. At Zacango, educating children is an important aspect of the zoo's mission. "Despite having survived dinosaurs, volcanic activity, and the crashing of meteorites, turtles find themselves in danger—in real danger—of extinction." He cites the usual factors—habitat loss, pollution, and climate change—and adds illegal trafficking fueled by the pet trade's appetite for exotic examples of nature's diminishing variety. "Trafficking exists because there is demand." But if turtles are not charismatic animals, why the demand? Pragmatism, says the vet. "Some people think that these are animals that require very little care and little maintenance." That's the turtle-as-pet crowd. But other types of turtle "owners" are driven by obsessive and perverse greed. "If I find out there are only twenty turtles of a particular kind that are wild and living," says Dr. Álvarez, "I will want one of those twenty. These people want to show off their collections. They will show off endangered turtles, knowing there are only twenty, and brag about having three."

Álvarez and his colleagues have acquired much experience caring for sick turtles. "Many people come here with turtles that are in very bad shape—malnourished, sick with parasites, with very soft shells because of bad diets. If people have dogs, they take them at least once a year for a checkup. But they think a checkup is not needed for turtles." We walk through the *herpetario*, past exhibits of lounging reptiles—snakes, lizards, iguanas, and turtles—to a zone that is off-limits to the public; there, we sanitize our shoes and walk into the room that is serving as an ad hoc clinic for the sick and damaged turtles saved from the sinister Mexico City apartment. Now at Zacango their recovery progresses: clean water, good food, medicine, and a hopeful prognosis.

Crucial work, Dr. Álvarez calls the turtle rehab we're witnessing. But why? I ask. Why the extraordinary effort to save twenty turtles, especially when Mexico City is teeming with millions of people living in deplorable conditions?

— *Turtles as Existential Tutors* —

Álvarez's answer is both heartwarming and sobering. When animals come to his zoo for help, says the vet, "we don't care if they are on the verge of extinction or whether they are beautiful. A life is a life. A life is the same to a bee, to a tarantula, to a serpent, to a turtle. The principal thing to each one of those beings is their life. Life we cannot give back. If a boa dies, to say, 'There are many more,' is to agree with Nazi concentration camps because we humans are many. There are so many of us on the planet, and if we needed to cut the number, who would say, 'Me, me! I volunteer to die!' Nobody would."

Whew. From sick and injured turtles to Nazis to existential questions in one breath. But the vet continues to make his holistic point. "That is why it is so important for us that children have the opportunity to have in their hands a snake or a turtle. A child who is fascinated because they had a turtle or a serpent in their hands will have a difficult time killing an animal. Such a child will not harm them." As he explains the importance of saving the twenty turtles in his care, Dr. Álvarez rails against mass extinction, humanity's devastation of Earth, the irresponsible and unsustainable use of natural resources. "I appreciate being able to sit on the top of a mountain, in a forest, in the jungle, and to be in silence and only listen and see." He's speaking like a preacher. "I have had the opportunity to see a jaguar and a tapir in the jungle, and it was a religious experience. Why take away that experience for tomorrow? What if my children or grandchildren or yours wanted to do the same and there are no forests or jungles, no jaguars or turtles?" We climb into the car for the drive on the jammed highway back to one of the most populated and most polluted cities in the world, a pulsating example of his passionate worries.

Whether the Mexico City turtle abuser was trafficking in turtles or hoarding them for his own amusement was not clear to investigators when they searched his apartment. But Claudia Edwards sees hoarders abusing animals with their obsessive collecting habits. "Such people are not connected with other people," she theorizes. "It's loneliness. They're isolated. And animals don't judge them."

That's understandable: companionship that doesn't talk back. But why not just one or two turtles? Why a score of them?

"I think because if one turtle makes you happy, you'll think two will make you happier," says Dr. Edwards. After that it's easy to rationalize adding more—more turtles equals more happiness. But there's another factor. Caring for the menagerie requires attention, and such attention takes the lonely and isolated mind off the hoarder's personal problems. Perhaps the initial motivation is a valid one: providing a home for a rescue, for example. But as the hoarder collects more and more turtles, caring for them becomes a new issue for the collector. "It's a psychological problem," says Edwards. We're fighting the traffic, so we have plenty of time to talk turtles. "They lose the ability to recognize the animals' pain. They only think of their own pain." But the turtles she rescued were damaged, injured, and diseased. How could such abuse be ignored? "They don't see it," she says about hoarders, because they are sick. Of course traffickers may see it but not care because profit is the goal.

Back in Mexico City I stopped off at the National Museum of Anthropology and checked out the Mayan images of turtles, images of carapaces with humanlike heads and arms. Turtles in Mexico through the ages: from the pre-Columbian clay deities to the capital city's old streetcars—so slow they were nicknamed *tortugas*,[1] and the score of injured chelonians recovering at the Zacango zoo.

• Cute Fred •

Fred looks so sweet, and so cute: he's resting under his basking lamp, both his front legs stretched straight out, his neck extended, with his little head resting on his left foot. But when he looks at me, I can't (of course) determine if he's saying "Hello" or "Don't bother me" or "Bring me another night crawler." I do so admire his patience—and I appreciate the patience he's teaching me. Watching his stress-free basking invariably slows down my frenetic usual pace—a least for a few moments at a time. Thanks again, Fred.

Time for some grub. I mix some shredded lettuce (romaine!) and diced apple in with Fred's dog food and serve it to him in the plastic container—it's a little cup—in which the dog food is packaged and sold. The cup is about a half-inch deep, so it's an awkward stretch for Fred's neck as he lunges at the breakfast. No problem. He pushes at it until he manages to shove the container up against the wall of his house, dumping some of the mixture out where it's easy to grab, and making the rest of it easily accessible as he pushes his face into it, holding the plastic cup steady with his feet. I get my oatmeal mixed with walnuts and cranberry sauce and sit next to Fred, and we down our respective gruels together. Unlike Fred, I manage to keep my hands out of my food and my oatmeal off my face.

The cup is empty, and he looks up at me with . . . what? A request for more? A thank-you? Just curiosity? I tell him there's more on the *New York Times* travel section that's lining his house floor, back where he spilled some when he pushed the cup onto the wall. And darned if he doesn't rotate 180 degrees—moving like a bulldozer with its treads rolling in opposite directions for a tight turn—and lumber back to eat the remaining mess. Did he smell it? Remember he left it? See it with his wandering red reptilian eyes? Despite our growing closeness (at least on my side of the relationship), I'm sure not ready to believe he understood my sleepy English when I called attention to the leftovers.

His basking lamp timer switched on during breakfast. Fred ambles over to it, plops down on a few remaining shreds of romaine and pieces of apple, stretches out his neck, raises his head toward the heat, and yawns.

— The Uphill Fight —

Claudia Edwards' professional colleagues around the world ply their Sisyphean tasks. The South China Nature Society and other Asian conservation movements struggle valiantly against the overwhelming odds facing turtle populations. Often-contradictory forces are at play as turtles struggle to survive human predators: The do-gooder animal advocates parachuting in to ecotourism opportunities as voyeurs or volunteer helpers protecting endangered turtles. The professional conservationists. The game wardens. The police worldwide—some of them in cahoots with the traffickers. The hard-core poachers, smugglers, and black market merchants who don't hesitate to use violence to guard their looted prizes. The dirt-poor hunters and gatherers trying to feed their families by grabbing an endangered turtle here and some threatened turtle eggs there. The collectors who seek rare species in the belief that consuming parts of the animals will cure them or that keeping such animals as "pets" will satisfy their lust for power and the forbidden. All of these forces have separate motivations, often at odds with one another. And all of them have repercussions on human lives, intertwined with habitat degradation and loss.

A nasty example belongs to the forty-fifth president of the United States. In the late 1990s he broke ground for the Trump International Golf Club in West Palm Beach. It was ground already well-broken by the burrowing, threatened, and protected gopher tortoise (*Gopherus polyphemus*). "They had to be cared for, absolutely," Trump proclaimed. "We were entering their turf, and we wanted to make sure we found an equal or better environment for them. Safely relocating them became a priority. I learned a lot about gopher tortoises. For example, they have been known to dig burrows as long as 40 feet by 10 feet deep. Just imagine what that could do to a golf course! So while I admired the tortoises for their industry, they had to be carefully relocated."[2] In the midst of the 2016 presidential election campaign *Fusion* investigative reporter Deirdra Funcheon dug into the story. She learned that Trump added nine holes to the course in 2006; a year later, when a state-mandated tortoise census was conducted, inspectors found no gopher tortoises. What happened to them? "It's unlikely they could have migrated very far on the course without drawing attention," Funcheon reported. "It's less likely they would have passed beyond the course's fenced boundaries. If they died, how? And what happened to their bodies?" No one in the Trump Organization answered her queries.[3]

In Madagascar, adjacent to a government ploughshare tortoise preserve, there is a Chinese-financed and -operated iron ore mine. The previously isolated ploughshare habitat is now easily accessible from the mine's newly constructed road network, which links to the closest port. This creates a ready trade route for smugglers—in this case, workers traveling back and forth from the Chinese mainland to the Madagascar job site. It is a crisis replicated worldwide at infrastructure projects, especially in Africa: the marriage of legitimate business deals and the parallel plundering of turtles.

Who is to decide between sustaining species and sustaining impoverished families? What is more important—saving a few endangered olive ridley eggs and ploughshare tortoises or poaching them to feed children? This is a daily question in the third world, one that first-world jet-setters who advocate conservation or enjoy ecotourism never face.

• Basking Fred •

Fred is doing his basking thing big-time. He is directly under his lamp, his front legs stretched out, his neck stretched out between them with his head resting languidly on the *New York Times* arts section that's spread over his heating pad, and his back legs stretched out on both sides of his tail. Fred must be relaxed. I think this is the longest I've ever seen his tail stretched (and why should I think a stretched-out tail is a marker for relaxed?). He looks like a bathing beauty at the beach. No, that's not true. That's an anthropomorphic leap that defeats even my own Fred-loving credulity. He looks like a turtle stretched out on an old newspaper in a box. Sheila rejects my calling his place a house. She calls it a cage and says a desert box turtle should be roaming around the desert—free. That's one thing Fred is not when he's living with us: a free-range turtle.

But he is well fed. This basking follows a heaping tablespoon of Whole Paws brand ("exclusively at Whole Foods Market," preens the label) grain-free chicken dinner with added vitamins and minerals. No wheat (in case Fred should be on a gluten-free diet?), corn, or soy. "Love is a four-legged word," reads the yuppie messaging on the dog food package.

I realize, as I watch Fred gobble the tablespoon full of Whole Paws, that I may not know if Fred is happy, but I do know that it makes me happy to watch him eat. Although Matt promised when he FedExed Fred to me that caring for him was idiot-proof, I worry plenty about good ole Fred. A "Feed Fred," reminder written on a Post-it note is stuck to my laptop. Sheila is worse. She grates carrots, minces persimmons from our neighbor's tree, mixes in bananas, and stirs it into the dog food, "so he gets all the stuff he needs," she explains, "that's not in the dog food."

Fred, as laid-back as he acts, is plenty of work. He doesn't leave neat droppings like hares, nor does he drain his bladder in an easy-to-clean cat box. He does what he needs to do when and where it happens to happen, as best I can map his powder room activities. And that makes a mess of the box.

I know I'm going to miss hanging out with ole Fred when he heads back to Arizona, but I also know I'll breathe a proverbial sigh of relief when he's back in Matt's experienced and turtle-loving but pragmatic hands. I imagine Matt will just toss leftover dinner in Fred's general direction, and maybe that's better for Fred than our fussy gourmet offerings. Or at least as good and without the bother.

ELEVEN

THE IMMINENT FUTURE

— Global Warming and Abandoned Turtles —

Richard Branson is not the only turtle fancier in Britain, but only he can boast of having a whole island in the Caribbean for his dole of turtles and tortoises. There he works with friends and colleagues to save turtles. "There's a wonderful feeling that comes with being able to help an endangered species survive another day," he says about helping rescue endangered sea turtles. "Releasing healthy wild animals back into their natural habitat is a humane way to help individual creatures. But there's an even better feeling that comes from working to prevent their extinction all together."

A group of turtles is sometimes called a herd or a bale, other times a nest or a turn or a dole. But turtles have been called much worse in the British Isles: predators and pests. The problem is the pet trade once again—even the legal pet trade. Mom and Dad might mean the best when they bring home a turtle for their tots, but too often the turtles become a burden and they're dumped in accommodating canals and ponds. "When these animals are bought as babies they seem attractive pets," says John Baker of Britain's Amphibian and Reptile Conservation Trust. He understands the initial appeal of almost any baby animal, "but they grow to a significant size and people think it is okay to take them to their nearest body of water and release them into places where they prey on native species and can spread disease."[1]

Not to worry too much, thought experts, when turtles first started to plague the UK. At least the critters won't multiply, they reckoned, since Britain's winters are much too cold to allow chelonians to breed and lay eggs, let alone for those eggs to hatch and for the hatchlings to survive. Enter climate change and global warming. Turtles *are* breeding and eggs *are* hatching. Ponds and canals are filling with British-born turtle subjects.

The result: decimated local wildlife, from newts, fish, toads, and insect larvae to ducklings and little coots.

The problem is not limited to the legal pet trade. Britain, despite its reputation for tough border controls, is not immune to turtle smuggling. At Manchester Airport in 2008, to pick just one port of entry and one random year, customs agents grabbed eighty-seven painted black turtles inbound from the States; a month before, more than fifty were found transiting the same airport.[2] Keepers at the London Zoo, like Rachel Jones, lament their fate. "There's a huge trade in reptiles. Tortoises and turtles are often confiscated," she says, "and it's extremely difficult trying to find homes for these animals."[3] Even Her Majesty is afflicted. Among the gifts she's been forced to accept as queen are live animals: for example, two sea turtles from the Seychelles. It was off to the London Zoo with them, though—there simply was no place for them in the queen's household, even in sprawling Buckingham Palace.[4]

— *I'll Not Abandon You, Fred* —

My musings about Fred bring me to Gilbert White and his exhaustive 1789 natural history of the English village where he lived his life, Selborne. Figuring prominently in his study was his adopted turtle, Timothy. White contemplates hibernation and wonders why Timothy "squanders more than two thirds of its existence in a joyless stupor." White attributes Timothy's lazy routine to his shell. "Pitiable seems the condition of this poor embarrassed reptile: to be cased in a suit of ponderous armour, which he cannot lay aside; to be imprisoned within his own shell, must preclude, we should suppose, all activity and disposition for enterprize." But White notices that Timothy manages to get out of the garden at the beginning of summer. "The motives that impel him to undertake these rambles seem to be of the amorous kind: his fancy then becomes intent on sexual attachments, which transport him beyond his usual gravity, and induce him to forget for a time his ordinary solemn deportment."[5] Gilbert White may well have identified with Timothy's solemn deportment. He lived alone. Timothy, herpetologists determined post mortem, was female.

What I won't consider is letting Fred just wander out of my backyard toward the Willamette River. When my travel schedule precludes me from taking proper care of Fred, I won't dump him at Delta Ponds just hoping he survives. I'll call Ship Your Reptiles.

— *Cuban Turtles Outlive Fidel* —

Soon after Raúl Castro and Barack Obama agreed to resume diplomatic relations between Cuba and the Yankee imperialists, I was on the island to check on rumors that Raúl's older brother, Fidel, fancied turtles and presided over his own secret turtle reserve at his swanky private resort island.

Fidel and his relationship with turtles stimulate the type of oblique jokes told both for laughs and as social commentary. Take Fidel's longevity as Cuba's leader, for example. A generic rally celebrating the revolution is the locale for one of these jokes, and there a Fidel fan gives *El Jefe* a Galápagos tortoise.

"How long will he live?" Fidel asks his follower.

"A long time," he is told. "As much as a hundred and fifty years!"

When he hears this, Fidel rejects the gift, telling his fancier, "No, *gracias*. That's the trouble with pets. You get attached to them and then they die on you."

Little Havana, the epicenter of the South Florida Cuban community, is a tourist attraction. Busloads of camera-toting visitors stop off for a taste of Cuba à la Florida, from the strong and sweet espresso to the Buena Vista Social Club–sounding music to the aging Castro refugees playing chess and dominos in the city park reserved for their games. Yet Calle Ocho—Little Havana's main drag—remains *auténtico*, and it was there, at the Versailles ("The World's Most Famous Cuban Restaurant") that I met with Fernando Bretos for a quick tutorial on the status of Cuban chelonians.

Ruddy-faced marine biologist Bretos looks beached in the citified Versailles bakery. His couple of days' stubble makes him another Indiana Jones look-alike as he rhapsodizes about the healthy Cuban sea turtle colonies. We're drinking caffeine-laden *cafecitos*, and we're talking fast; both of us are on tight schedules, heading soon for Cuba. It's a place where Bretos has been working since 1998 and a place where he feels at home. His parents were "Peter Pan" immigrants—two of the thousands of children who were relocated to the United States in the early 1960s by families who did not want them growing up in a Castro-controlled Cuba.

Bretos is the director of the Cuba Marine Research and Conservation Program, a project of the Ocean Foundation, and curator of ecology at the Patricia and Phillip Frost Museum of Science in Miami. He coordinates U.S. sea turtle research with colleagues doing similar work at the University of Havana, and he works with rural coastal Cuban communities to help them protect rather than plunder natural resources. Since 2000 he's been monitoring the three species of sea turtles—loggerheads,

hawksbills and greens—that nest on Cuban beaches, and he reports growing populations.

— *Cuba as a Refuge* —

The Cuban government officially stopped the harvesting of turtles for food in 2008, after inspectors working to enforce CITES
regulations proved that illegal Cuban turtle shells were showing up in the clandestine Japanese market. Unlike other Caribbean and Central American turtle predators, "Cubans don't
touch the eggs," says Bretos. "We've found shells with eggs in
them. They eat the meat." They ate the meat and sold the
shells to supplement their diets and incomes. But Cuban law
slammed the brakes on the turtle trade and since the 2008
change in official policy, Cuban-nesting sea turtles are thriving.

"Cuba's done well for turtles with three millions tourists,"
Bretos tells me, draining his *cafecito*. For several decades, most
Americans were barred from visiting Cuba by the U.S. embargo;
however, Europeans and Canadians have flocked to Cuba
throughout this period, but in comparatively modest numbers.
With the Castro-Obama embrace, Bretos now worries about the
turtles' futures when hordes of Yankees will again descend on
the island for rum and fun, cigars and guitars.

"What will happen with more tourists?" he asks, his animated
face alive with a conflicting combination of energy, smiles, and
worry. Florida, he points out, hosts almost 100 million tourists
a year, Cuba only those few million. "Raúl has no choice but to
open Cuba to Americans," he says, since the Castro brothers lost
their economic lifelines when first the USSR and then Venezu-
ela stopped subsidizing the country. "I don't think Americans
are going to want to eat turtle. Maybe a bit, but not enough to
make an impact. I think the turtles will end up okay." He's more
concerned about overfishing and lobster trapping that's not

sustainable. "Turtles will do okay. It's the other guys . . ." He thinks about vast hotel resorts that could invade pastoral Cuba and about their impact on the environment. But even consequential new developments need not devastate turtle habitat in Cuba, he says, because of its extensive coastline.

Before we part, Fernando Bretos insists I meet Félix Moncada once I arrive in Havana. When CITES pressured the Cuban government to stop harvesting turtles for food, it was Dr. Moncada, coordinator of the sea turtle program at the Cuban Fisheries Ministry Research Center, who oversaw the closure of the fisheries.

— *Cuba Turtle Lore* —

What to read traveling around Cuba in search of turtles? The journals of Columbus, perhaps, who wrote that he observed turtles in Cuban waters "in such vast numbers that they covered the sea." Or a letter written home by his crewman, Michele da Cuneo, who reported finding "huge, infinite and excellent to eat" turtles on Cuban beaches. Maybe the critical analyses of colonialism by the reformist Spanish priest, Bartolomé de las Casas who learned that the indigenous population kept hundreds of sea turtles fenced in the water and ready to eat near present-day Cienfuegos (remarkably close to Castro's private island, Cayo Piedra, and his own supposed turtle stash).

Hemingway? He drank daiquiris at what's now a Havana tourist trap, the Floridita bar, and wrote *The Old Man and the Sea* while living in Cuba. "Most people were heartless about turtles because a turtle's heart will beat for hours after it has been cut up and butchered. . . . I have such a heart too," muses the old man.

I chose a lighter route, Graham Greene's self-described "entertainment," *Our Man in Havana*, published just as the Batista dictatorship was succumbing to Castro's guerilla war. Its protagonist, a vacuum cleaner salesman turned spy, is inspired by interlocking Hoover canisters and hoses to submit reports to his London handlers about imaginary rebel weapons. It, too, is a book with chelonian references, as when an ex-pat doctor appears with "seamed and sanguine skin [that] could change no more than a tortoise's" and a competing electric cleaner dealer takes "a swig of turtle soup." Greene proves to be a prescient guide for Cuba's future and for my own quest for Castro's turtles.

The lobby of the Havana Libre hotel seems much like it did when Graham Greene was wandering around 1950s Cuba except for the wall of black-and-white photographs celebrating the arrival of Fidel and company. Once they conquered Havana, they famously encamped in what was then the Hilton, with their trademark beards and fatigues, leaning their rifles on the overstuffed couches. As I waited to meet Dr. Moncada, the hotel driveway took on a periodic late 1950s surreal look as Havana-cliché old American cars, hired for joyrides along the seaside Malecón, disgorged giggling tourists.

Félix Moncada strides across the old Hilton lobby to the bar where I'm nursing a drink. He's wearing a wide relaxed grin and a T-shirt festooned with the image of a turtle. Prehistoric evidence, the government's turtle expert informs me, shows that sea turtles were always used by the island's inhabitants—for food and for their skins and shells. Until the revolution, he says, the turtle population was not negatively impacted by the harvest because Cubans hunted with primitive boats and the turtles they caught mostly served only a traditional local market. After 1959, the government deployed more sophisticated fishing boats to the turtle fisheries and encouraged development of the sea turtles as a nationwide food source. By the late 1960s, he remembers, laws were enacted to ban turtle hunting during egg-laying season and hatcheries were established to protect eggs from poachers and predators. Still, says Moncada, because the turtle grounds were so remote from the cities, Cuban-based sea turtles never faced an existential crisis.

— *Turtle Tourism* —

Cuba now sees the value of its protected sea turtles as a draw for ecotourism. Cayo Largo del Sur is an example. Tourists can visit a turtle farm on the island and spend summer nights watching for turtles to come ashore and lay their eggs.

Meanwhile Félix Moncado knows that no matter what the law demands, some turtles will succumb to poaching. He understands the appeal. "I have eaten turtle, but I haven't eaten it for a long time, to set the example and as a matter of principle, because it is illegal. While I was in the boats at the sites studying them and implementing measures to protect them, the fishermen would get turtles and sometimes some of them ended up dead. I tasted them." How do they taste? "Delicious," he reports with candor. "The best is the green turtle because it tastes like a beef steak." Other species he dismisses as greasy or tasting like seafood. He knows bycatch turtles supplement the diet of hungry fishermen, and he understands why—Cuba's economy is a mess and it's challenging for most Cubans to find a regular supply of high-quality foodstuffs. "It is still difficult to obtain meat, so people, if they have the possibility to eat a turtle steak, they will eat it. If they do not have beef they know that the turtle is an alternative. That is why we still have illegal fishing. Some unscrupulous people here are like people everywhere else. Some predators cast their nets to get turtles. However, we do not have people who go to the beaches searching for their eggs. That is practically nonexistent here. But we do have some people who live in Havana near the seashore, and they know the months when the turtles are available. They cast their nets to get them and sell the meat." Maybe, he says, the buyers just want to brag that they ate a turtle steak. One thing he knows for sure is that the meat fetches a hefty price.

— *Turtle Metaphors* —

Turtles do not just populate the UNESCO Biosphere Reserves in Cuba; they inhabit the home of UNESCO's Cuba station chief, Fernando Brugman. We're sitting in his office in the former swank headquarters of a Spanish colonialist, a settler who clearly suffered from too much money and from taste in interior décor that could best be described as cultural indigestion. Faux Grecian columns compete with Crayola 64–colored ceramic tiling in just one example of a roomful of excess. Nonetheless, the building is impeccably restored to its glitzy glory, a sharp and refreshing contrast to the crumbling ruins that pockmark so much of Havana.

"You keep turtles," I prompt Brugman. I've been told that he is a fancier.

"Yes," he acknowledges. "First I had Bureaucracia. Bureaucracy," he translates. "I bought her for one dollar in the countryside around Havana." She's big, he reports, the size of a dinner plate. And she's not the first turtle with that name. The Argentine cartoonist best known as Quino gave his star character, the six-year-old girl Mafalda, a pet turtle she called Bureaucracia, so named because it seems to move in slow motion.

"What type of turtle is Bureaucracia?" I ask Brugman.

"I don't know," he dismisses the questions with a pragmatic answer. "It's a Cuban turtle."

He saved her from dinner or from Santeria, "or for using it for something else. They have these religions and rituals," he muses, "in which they use certain animals. I prefer not to see those types of things. But I like turtles, so I got a turtle."

"Welcome to the Cuban jungle. This is how we survive," Elias Aseff, a cultural studies academic and Santeria practitioner, tells me in response to Brugman's reluctance. Chomping his ever-present cigar, he's touring me through another Santeria house. "Rats, mice, and chickens are all sacrificed."

I look at a wooden turtle on what appears to be an altar. "And turtles," he adds. "We are allowed to sacrifice animals by Cuban law because this is part of our identity. It's normal and not forbidden. This is not the USA," he puffs. "This is Cuba. Our points of view are different. Enjoy my country," says a contented-looking Aseff, "just don't try to understand it." It is a mantra, motto, and excuse repeated throughout Cuba, often announced with great pride.

Despite the "just don't try to understand it" assignment, a pedicab driver I meet in Habana Vieja helps me learn how Cubans exist when so many official salaries are about $20 a month. I'm wandering the tourist district and he's waiting for a fare as we chat. He introduces me to the cabbie next in line and tells me, "He works in a cigar factory."

"Want to buy a *habano*?" asks the second cabbie, who uses his access to the factory's stock as a convenient device to augment his income.

— *Castro's Secret Turtles* —

As the lazy Caribbean sunny days lull me into a slow Latin rhythm, I'm soaking up Cuban realities mixed with that rumor I heard back in New York of a private Fidel turtle reserve. I'd traced it to a kiss-and-tell book written by Juan Reinaldo Sánchez, who claims to be a former Fidel bodyguard. Now living in exile in Miami, Sánchez worked with French journalist Axel Gyldén, a reporter at *L'Express*, on the book, which was published in English as *The Double Life of Fidel Castro*. Sánchez details what he claims was Castro's life of luxury on Cayo Piedra, his private island just south of the Bay of Pigs. Gyldén informed me that Castro did indeed keep turtles on Cayo Piedra—but not because he was acting as a sea turtle protector and conservationist. The turtles were kept in offshore pens much like the Cayman Islanders made when they first arrived on the southern Cuban shores two hundred years ago, and for the same purpose.

"On his private island, Castro had a small turtle farm," Gyldén told me when I reached him at his Paris office. "They were for his own consumption!" The journalist sounds exuberant as he rejects the idea that Castro was providing a refuge for endangered species or a private zoo for his own enjoyment of frolicking rare turtles. "He was eating them," he reports. While we talk Gyldén tells me he's sending an email to Sánchez to check the facts. "Of course Castro ate the turtles," comes the confirmation from Miami. "It was just easier to keep them at home, a privilege only a king has." Sánchez isn't alone in commenting on Fidel's diet. Back in 1998, *Le Monde* reported that Fidel liked "plain nourishing food, especially turtle soup," a quote the newspaper attributed to Castro's fellow revolutionary (and rumored lover), Celia Sánchez.

Maybe it's the Santeria. Maybe it's the Havana Club rum. Maybe it's *Our Man in Havana* and his vacuum cleaner–inspired weaponry. But I'm starting to see what look to me like giant turtles and tortoises rolling down the Havana streets. After a few days in Cuba, the flow of old American cars is not much of a novelty; it becomes the kind of thing current jargon calls the new normal. Some of the most common models still going strong in Cuba are the 1950, 1951, and 1952 four-door Chevrolets. The more I watch them cruise the rutted streets, the more they look like chelonians to me—their rounded hoods and trunks curving up into their roofs seem so akin to the highly arched carapaces of ploughshares. The two-tone Chevys look like a fancier species—maybe Burmese stars, especially, for some reason, those Chevys with light blue bodies and off-white rooftops. I try my hallucination out on a *taxista* who is piloting me in his pristine early-1950s all-black Chevrolet sedan. "*Sí*," he agrees. "We call the '49s 'torpedoes' and, now that I think about it, mine does look like a turtle. *Torpedos y tortugas,*" he considers this new image for his ride and likes the idea.

Again Cole Porter asks his apropos signature question for us all, "Is it the good turtle soup or merely the mock?" I know now, after living with him, that Fred is good. We've become friends. Or at least he's a friend of mine. I like to think I take good care of him, that he's happy in his cozy enclosure. I hope he's not lonely and that I provide some semblance of good company for him. We commune daily, or at least I talk to him and watch him and try to learn from him.

One thing I've learned from him—and from this turtle and tortoise quest I've been on—is that saving chelonians is of existential importance to us all. They are canaries in the coal mine called Earth where we all live. Even if we humans force the demise of these animals connected to ages that predate us, our disdain for their environment translates by definition to a lack of adequate concern for the sanctity of our own. So this quest has become a call to action.

A CALL TO ACTION

Why don't we tend to anthropomorphize reptiles? Probably because we don't see ourselves in snakes and lizards and turtles. They don't seem to look like us. A polar bear's face is not all that different in structure from our own. A kangaroo joey hanging out in its mama's pouch in the Outback doesn't look much different than a tyke in its Snugli on a stroll through Central Park with its mother. Of course a penguin is just one of us in a tuxedo. But a snake? Where are the arms and legs? A lizard? Its eyelids are upside down and it's covered with scales. Snakes and lizards are not us. As for turtles, with their wrinkled and jowly heads hiding under a hard shell, all but helpless when they're flipped over on their backs, they're strange beasts and hard to consider as cousins.

Yet cousins—very distant cousins—they are and we must stop the human predation that's devastating turtle populations worldwide. I recall a graffito scratched onto the control panel of an elevator in a San Francisco office building back in the 1970s, a line that was intended as a sardonic commentary on political correctness: "Save the gay homeless whales." Silly, but the underlying message rings true.

Consider the imminent extinction faced by scores of turtle species, what that means for the health of the planet and what courses of action might reverse the collapse of the venerated Yangtze giant softshell and the lumbering ploughshare along with those common box, mud, and painted turtles being vacuumed out of the Louisiana swampland.

• **Fred Separation Anxiety** •

So I'm feeling lonesome for Fred already even though he's upstairs in his house in the living room. Because I just sent this email message:

> Good evening, Matt.
> Fred sends kisses. He was a hit last week at the university. I took him in his official Tortoise House to visit my students. He has been fascinating to live with. You and my Cuban Santeria priest were correct. It was a requirement that I live with a turtle. But now I would like to take advantage of the option you offered and send him back home to Arizona. It is not that we've failed to bond. On the contrary. He seems to like me, even if I cannot teach him to fetch the newspaper and my slippers, and I like him. But yesterday I looked at my upcoming extensive travel schedule over the next six months, along with the options for his care and feeding while I am on the road, and nothing seems appropriate other than a return trip for him to your place. Hence this note to check if he's still welcome to come back and, if so, a request for instructions on his travel arrangements. I imagine it is simply to reverse the procedure that got him here? Again, thank you very much for allowing us to foster Fred and I look forward to the opportunity to come down to Arizona sometime and visit him along with you and the rest of the menagerie.
> Cheers, Peter

Matt wrote back within the hour:

> Peter,
> Please look at ShipYourReptiles.com. They short-cut the process. You can call me anytime to explain the prep. The process and finances are handled by the website. I hope you remember how he was shipped to you. If not, I can walk you through it. The bigger thing is that I am very, very glad that he afforded you and Sheila an opportunity to bond with a primitive being. They provide us with a different view of the world and our relationship with it. Think slow and steady. Unfortunately, in today's rapidly environmentally changing world, slow and steady may not be enough, and these archaic species find themselves lost in the changes. I am a turtle and tortoise lover because I have always loved the underdog, human and otherwise. These noncharismatic species *are* the canaries in our environmental coal mine. As they go, I believe, we go. I have a favorite graffito I like to write in bathrooms (with chalkboards): Vermin Rule!!! My version of the Meek Shall Inherit the Earth. Enough of my lecturing. Fred is always welcome here.
> Best, Matt

His welcome lecture prompted me to respond:

> And you are always welcome here, Matt. Yes, we just fed Fred one of his Walmart night crawlers, and watching him slowly stretch out his neck, turning his head to find what he sensed was waiting for him (I guess my saying, "There's a worm over by your left hand, Fred," was not his cue), and then lunging at it, grabbing it, wrangling it, biting it into pieces, and chewing it down while those pieces still writhe is a primal experience that forces me to

think about more than university politics or traffic or even Trump. I know that living with Fred broadened my worldview. I know that watching him slows down my tendency toward nonstop work and play. I can watch him without distraction and without losing interest in him even if all he's doing (as is often the case) is basking. Looking at the crinkles and wrinkles on his extended neck, watching it articulate when he moves his head, seeing where the leathery skin disappears into its mysterious shell casing, trying to figure out what motivates him when he changes his lounge position or chooses to saunter into his dark room or flop into his water bowl. That saunter so fluid and yet primitive, like an early Charlie Chaplin.

Again, thanks so much, Peter

I log on to ShipYourReptiles.com and fill out the FedEx forms for a departure tomorrow, realizing I'm already suffering separation pangs.

It's easy to be preoccupied with daily routines and hassles; it's easy to ignore a slow-walking reptile that's been quietly minding its own business for millennia and without a close look appears healthy and happy. Our concern cannot be mock. Hence, take a close look, heed the warnings, and work on a blueprint for change. Before it's too late. Now.

• Adios, Fred •

It's snowing this morning when I go out to the driveway to fetch my *New York Times*—maybe that's a good sign. Fred is going home to Arizona today, shipping out of the wintery Pacific Northwest. We look at each other, and I give him a couple of slices of my breakfast banana. I've retrieved the cardboard carton he traveled to Oregon in, along with the cotton sack he was wrapped up in and the Styrofoam peanuts that cushioned the plastic box that served as his overnight cabin.

Blood from yesterday's worm treat is dried on Fred's *New York*

Times hot pad cover, adding to the *Jurassic-Park*-in-my-living-room atmosphere he's brought into my life. I'm realizing that when I look at him, I'm connecting through him—even if I'm forcing the mood—back those 200 million-plus years to the dinosaurs. The nails or claws on his feet or paws look as prehistoric as they are, as does the at-rest position of his front legs: bent at the elbow, his feet soles up. My non-scientific and inconsistent references to his physicality only add to the anthropomorphizing I foist on poor, innocent Fred—maybe I should have called him Turtle Specimen Exhibit A and referred to his body parts with the names used by herpetologists, a preventative tactic to getting too attached to him. So it goes. Out he comes from his house for a photo op and then it's into the bag, into the box, and a quick ride in the old Volvo down to FedEx.

Bye, at least for now, Fred.

At the FedEx shipping center Beverly looks at the box and the Sharpie-marked words, LIVE ANIMAL. "Whatcha got in there?" she asks. "Lizards, snakes?"

"Fred," I tell her, "he's a box turtle."

Beverly melts. "I had a box turtle in third grade," she tells me. "They're so sweet," and her face looks like a little girl's again. "I took him to school with me every day. In my lunch box."

Everybody has a turtle story.

And Fred lives happily ever after. FedEx delivered him alive and well back to Matt Frankel's Arizona oasis. There Matt and his team breed endangered turtles and tortoises, maintain assurance colonies, and reach out into the community to educate the public about our natural world. Fred plays a starring role, acting as a hands-on example of what we risk losing if we ignore threats to biodiversity. Good ole Fred.

ACKNOWLEDGMENTS

More or less in the order I encountered them as I researched and reported *Dreaming in Turtle*, I want to acknowledge a bale, this turn, my dole of friends, colleagues, and fellow turtle travelers for their generous help. Here I'll note those not cited in the text even while begging the pardon of those I might inadvertently omit.

At the Turtle Conservancy, board members John Mitchell, Russ Mittermeier, Anders Rhodin, and (old friend) Rick Ridgeway for their who's who of turtle subculture contacts. Turtle Conservancy journal *The Tortoise* editors Andrea Danese and Zoë Lescaze, researcher Kaitlyn-Elizabeth Foley, and art director Maximillian Maurer for research support.

University of Oregon School of Journalism and Communication (SOJC) Global Education Oregon (GEO) study abroad students Colin Brennan, Natalie Enger, Christian Hartwell, Morgan Kaufer, Natasha Kimmell-Harris, Kendra Siebert, Jessica Paulsen, Nathan Stevens, and Sabrina Zawierucha for surveying turtles on offer in Rosario, Argentina, pet stores. UO student Zoë (Ms. Monsanto) Haberkost for sharing her glyphosate research. UO faculty colleague Will Johnson for field research partnership in northern Argentina, and GEO

Rosario site director Maria Nelida de Juano for sharing contacts. UO students Levi Gittleman and Andy Field for research support in Jakarta, Indonesia. UO graduate student research assistants Audrey Black, Thomas Schmidt, Matt Eichner, Thipkanok (Ping) Wongphothiphan, and Christopher St. Louis for a wide variety of crucial tasks. Supportive deans at the SOJC Tim Gleason, Julianne Newton, and Juan-Carlos Molleda, along with Senior Associate Dean Leslie Steeves. Supportive UO professors, including Doug Blandy, David Frank, Damian Radcliffe, John Russial and the late Alex Tizon and Tom Wheeler. SOJC Associate Dean Ray Sykes and SOJC financial manager Josh Buetow for helping manage logistics. Wizards in the SOJC IT department Ryan Stasel, Cameron Shultz, Joseph Shea-Bianco and Louie Vidmar. UO Vice Provost for International Affairs Dennis Galvan, along with Eric Benjaminson, Francis Bivigou, and Maryne Dana at the UO Gabon-Oregon Center for options in Gabon.

Premier Travel agents Judy Bailor and Paige Dewey for getting me there and back.

Writers Jeff Kamen, Terry Phillips, Chris Slattery, and Tom Steinberg for critiquing work in progress. Poet André Spears for historical references.

Fixer Manuel Elier Perez for negotiating Cuban infrastructure. Taxista Maikel Luis Peña Reyes for piloting the '55 Oldsmobile. Jean Fuller Gest and her Cuba Up Close company for my Havana flat. Ocean Foundation program coordinator Katie Thompson and University of Havana professor Julia Ricardo for *ayuda* in Cuba. Yunnan University professor Qun Jin, along with her students Yifeng Lu and Rui Liu, for navigating China. Friends Hugo Speth and Ines Widman-Speth for *Hilfe* in Swabia. Writer and editor Tom Christensen for locating oracle bones and San Francisco Asian Art Museum China curatorial project assistant Jamie Chu for access to them. At her Samara Info Center travel agency in Costa Rica, Brenda Dragonne for sea turtle tour arrangements. Sonia De La Cruz for translation help. At the University of Costa Rica, Professor Carolina

Carazo for *pura vida* perspective. At the Forum Journalismus und Medien in Vienna, Daniela Kraus and Gunther Müller for introductions in Austria. Paco Colmenares of the Humane Society International in Mexico for surveillance footage, and Daniel Juarez, curator of the Maya room at the Mexico City National Museum of Anthropology for after-hours entrée to the Maya exhibit.

Diane Best for Mojave Desert accommodations at her Rattlesnake Mustang Ranch and Su Garfield for Louisiana contacts. Amity Bass and Jeff Boundy at the Louisiana Department of Wildlife and Fisheries for introductions to turtle fishers. At the University of Colorado, Professor Harsha Gangadharbatla for connections in Colorado. For access to NOAA's Galveston facilities, the agency's Rhonda O'Toole, and former Galveston resident Rick Williams for Texas turtle aphorisms. U.S. Fish and Wildlife Service special agent in charge Nicholas Chavez for access to agents. Marilyn Veltri, the public information officer at the Morgantown, West Virginia, federal prison for access to prisoners.

Special thanks to fellow journalist and author Alisa Roth.

At MacKenzie Wolf, agent extraordinaire Gillian MacKenzie. At St. Martin's Press, turtle-loving editor Hannah Braaten, marketing maven Allison Ziegler, publicist (and turtle aficionado) Hector DeJean, editorial assistant Nettie Finn, along with acquisitions editor Emily Angell and copy editors Martha Cameron and Eric Meyer. Plus a palette full of appreciation for the *Dreaming in Turtle* jacket design to artist Kerri Resnick.

I am privileged to hold the inaugural James Wallace Chair Professor in Journalism position at the University of Oregon, a professorship endowed by the *U.S. News & World Report* foreign correspondent and Oregon alumnus James Wallace and his wife, Haya. His datelines ranged from Moscow to Peking to Saigon, from Havana (Fidel Castro reportedly expelled him from Cuba for his reporting) to capitals throughout Latin America. Unfortunately I never had the opportunity to meet Wallace, but from everything I've learned about him, I wager we would have enjoyed each other's company—especially sitting at a saloon in Kunming or Libreville, Paris or Havana, nursing Bols gin and tonics, trading tales about being on the road in search of the next story.

And, of course, loving thanks to my wife Sheila, for everything.

NOTES

Chapter 1: The Majestic Turtle

1. Emily Thomas, "Survival Stories: Fish Bacon for Breakfast," *The Food Chain*, BBC World Service, aired August 20, 2016, www.bbc .co.uk/programmes/p044jjbx.

2. Walter Benjamin, *Charles Baudelaire: A Lyric Poet in the Era of High Capitalism* (London: New Left Books, 1973).

3. Clifford Geertz, *The Interpretation of Cultures* (New York: Basic Books, 1973).

4. Henry Nicholls, "Welcome Home, Lonesome George: Giant Tortoise Returns to Galapagos," *Guardian*, February 17, 2017, www .theguardian.com/science/animal-magic/2017/feb/17/welcome -home-lonesome-george-giant-tortoise-returns-to-galapagos.

5. Craig Sanford, *The Last Tortoise: A Tale of Extinction in Our Lifetime* (Cambridge, MA: Belknap Press of Harvard University Press, 2010).

6. Donald C. Jackson, *Life in a Shell* (Cambridge, MA: Harvard University Press, 2011).

7. Ingmar Werneburg et al., "Modeling Neck Mobility in Fossil Turtles," *Journal of Experimental Zoology Part B: Molecular and Developmental Evolution* 324, no. 3 (2015): 230–243, https://doi.org/10.1002 /jez.b.22557.

8. Jackson, *Life in a Shell*.

Chapter 2: The Timeless Allure

1. "Signs and Symbols: The Lyre," Fitzwilliam Museum, www .fitzmuseum.cam.ac.uk/pharos/collection_pages/18th_pages/M3 _1922/TXT_BR_SS-M3_1922.html.

2. Nga Pham, "Cụ Rùa: Vietnam Mourns Revered Hanoi Turtle." *BBC News*, January 20, 2016, www.bbc.com/news/world-asia-3535 8979.

3. Mike Ives, "Vietnam's Sacred Turtle Dies at an Awkward, Some Say Ominous, Time." *New York Times*, January 22, 2016, www.nytimes .com/2016/01/23/world/asia/vietnam-turtle-hoan-kiem-lake.html.

4. Shi-ping Gong et al., "Disappearance of Endangered Turtles within China's Nature Reserves," *Current Biology* 27, no. 5 (2017): R170–171, https//:doi.org/10.1016/j.cub.2017.01.039.

5. Fernando Bretos, email message to author, November 16, 2016.

6. "Caguama Beer," www.caguamabeer.com.

Chapter 3: The Voracious Consumers

1. Sean Sullivan, "The Greatest Quotes of Edwin Edwards," *Washington Post*, March 17, 2014. www.washingtonpost.com/news/the-fix /wp/2014/02/20/edwin-edwardss-greatest-hits-crooks-super-pacs -and-viagra/.

2. "Papoo's Turtle Soup," Stirrin' It Up with Chef John Folse and Company, www.jfolse.com/stirrin/recipes/recipes2012/SIU_101312 %20Papoo's%20Turtle%20Soup.html.

3. John Burnett, "Costa Rican Villagers Sell Turtle Eggs to Save Sea Turtles, but Feud with Scientists," Sustainable Development Reporting Project, http://lanic.utexas.edu/project/sdrp/tortugas .html.

4. "Dorismar Súper Modelo," WildCoast/CostaSalvaje, www.costa salvaje.com/who-we-are/heroes/12-dorsimar.

5. Stephen Calloway and Katherine Sorrell, *Obsessions: Collectors and Their Passions* (London: Mitchell Beazley, 2004).

6. Glenn Phelps and Steve Crabtree, "Worldwide, Median Household Income About $10,000," *Gallup News*, December 16, 2013, www .gallup.com/poll/166211/worldwide-median-household-income -000.aspx.

7. Arnold Zwicky, "Just Between Dr. Language and I," *Language Log*, August 7, 2005, http://itre.cis.upenn.edu/~myl/languagelog /archives/002386.html.

8. "Watch a Hummingbird at the Feeder, Fending Off All Other Hummers. The Rest of the Day . . . ?" *Pioneer Press*, August 15, 2012, www .twincities.com/2012/08/15/watch-a-hummingbird-at-the-feeder -fending-off-all-other-hummers-the-rest-of-the-day/.

9. Brian Christy, "Anson Wong Goes Free," *National Geographic Voices*, February 28, 2012. http://voices.nationalgeographic.com/2012/02/28 /anson-wong-goes-free/.

10. TRAFFIC Southeast Asia, "Anson Wong Appeal: Full Text of Statements by TRAFFIC Southeast Asia and WWF-Malaysia," Face-

book, February 24, 2012, www.facebook.com/notes/traffic-south
east-asia/anson-wong-appeal-full-text-of-statements-by-traffic
-southeast-asia-and-wwf-mala/360259414004909/.

11. Werner Muensterberger, *Collecting: An Unruly Passion: Psychological Perspectives* (Princeton, NJ: Princeton University Press, 1994).

Chapter 4: The Tawdry Marketplaces

1. "Turtle-Slicing Upsets Kids," *Abbotsford News*, December 20, 1997.

2. Shelagh MacDonald, "Inhumane Treatment of Turtles," *Animal Welfare in Focus*, Spring 1998.

3. Glen Clark, "Letter to Jim Harrington from the Office of British Columbia Premier Glen Clark," August 6, 1998, 78470-00/FWH-LINS01, Ref: MO 37420.

4. Glen Clark, "Sample Letter to Turtle Importers from the Regional Manager at the Fish, Wildlife and Habitat Protection Department of the British Columbia Ministry of the Environment, Land and Parks," May 5, 1998, 78470-45/FWH-LSST/NAME.

5. David Zuest, "Fax Notice to the Vancouver Humane Society from an Inspector at the Canadian Food Inspection Agency," May 30, 2000.

6. Marisa Lagos, "Ban on Live Turtle, Frog Sales Assailed," *SFGate*, May 6, 2010, www.sfgate.com/restaurants/article/Ban-on-live-turtle
-frog-sales-assailed-3265284.php.

7. *California Penal Code*, Title 14, Section 597.3.

8. Sam Pearson, "Asian American Leaders, Environmentalists Clash over Live Animal Imports," *California Watch*, February 24, 2011, http://californiawatch.org/dailyreport/asian-american-leaders
-environmentalists-clash-over-live-animal-imports-8873.

9. Riya Bhattacharjee and Stephanie Chuang, "Caught on Cam: SF Tour Guide Goes on Racist Rant," *NBC Bay Area*, www.nbcbayarea
.com/news/local/San-Francisco-Chinatown-Tour-Bus-Guide-Racist
-Rant-280204072.html.

10. "Our Story," *Commander's Palace*, www.commanderspalace.com
/our-story.

11. Oscar Wilde, "Dinners and Dishes (Review)." *Pall Mall Gazette*, March 7, 1885.

12. Alessandro Filippini, *The Delmonico Cookbook: How to Buy Food, How to Cook It, and How to Serve It* (Bedford, MA: Applewood Books, 2008).

13. Ruth Berolzheimer and Mary L. Wright, *The United States Regional Cookbook* (Chicago: Culinary Arts Institute, 1947), 275.

14. Lindsay Fendt, "Murdered Costa Rican Conservationist Had Been Chased by AK-47-Wielding Poachers," *Tico Times*, June 1, 2013,

www.ticotimes.net/2013/06/01/murdered-costarican-conservation
ist-had-been-chased-by-ak-47-wielding-poachers.

15. Lindsay Fendt, "4 Convicted, 3 Acquitted in Jairo Mora Murder Trial," *Tico Times*, March 29, 2016, www.ticotimes.net/2016/03
/29/4-convicted-jairo-mora-murder-trial.

16. Melissa Castellanos, "Costa Rica's Sea Turtle Eggs: A 'Killer
Aphrodisiac' Buried in Conflict and Cultural Traditions," *Latin Post*,
October 31, 2013, http://latinpost.com/articles/3444/20131031/costa
-rica-sea-turtle-eggs-leatherback-turtles-olive-ridley-turtles-black
-market-aphrodisiac.htm.

17. L. Arias, "Mob of Tourists at Costa Rica's Ostional Beach Prevents Sea Turtles from Nesting," *Tico Times*, September 9, 2015, www
.ticotimes.net/2015/09/09/crowd-tourists-costa-rica-prevent-sea
-turtles-nesting.

18. Jennie Erin Smith, "Murder on Moín Beach," *The Tortoise*, n.d.,
https://turtleconservancy/news/2016/3/murder-on-moin-beach.

19. "Policy Initiatives: International Issues: Cayman Island Turtle
Farm," *Sea Turtle Conservancy*, https://conserveturtles.org/policy-initia
tives-international-issues-cayman-island-turtle-farm/.

20. "Paul Backs Campaign to Stop Sea Turtle Farming," PaulMc-
Cartney.com, October 12, 2012, www.paulmccartney.com/news-blogs
/charity-blog/paul-backs-campaign-to-stop-sea-turtle-farming.

21. "Ending Sea Turtle Farming," *World Animal Protection USA*, www
.worldanimalprotection.us.org/ending-sea-turtle-farming.

22. "Cayman Island Turtle Encounters," *Cayman Turtle Centre*,
www.turtle.ky/explore/turtle-encounters.

23. "70% of Residents Not Eating Turtle," Cayman News Service,
October 9, 2015, https://caymannewsservice.com/2015/10/70-of
-residents-not-eating-turtle/.

Chapter 5: The Prodigious Farms

1. Li Yang, "Turtle Power Propels Qinzhou," *China Daily USA*, November 27, 2013, http://usa.chinadaily.com.cn/epaper/2013-11/27
/content_17135858.htm.

2. James Liu, "Rags to Reptiles: Building an Empire on One of
China's Rarest Turtles," *The Tortoise* 2, no. 2 (2017).

3. "Company Profile." Hoi Tin Tong Co. Ltd., www.hoitintong
.com.hk/about.php?lang=EN.

4. Amy Nip, "Power of Turtle Jelly Lies in Its Recipe: Hoi Tin Tong
Founder," *South China Morning Post*, September 18, 2013, www.scmp
.com/hong-kong/article/1312232/power-turtle-jelly-lies-in-its-recipe
-hoi-tin-tong-founder.

5. Rachel Nuwer, "Asia's Illegal Wildlife Trade Makes Tigers Farm-
to-Table Meal," *New York Times*, June 6, 2017.

6. Chiu Chun-Chin and Evelyn Kao, "Taiwan Seizes 3 'World's Most Expensive' Tortoises at Airport," *Focus Taiwan*, May 7, 2017, http://m.focustaiwan.tw/news/asoc/201705070016.aspx.

7. Emily Tsang, "Robbers Steal 12 Rare Turtles Worth HK$1.4m in Violent Pre-Dawn Break-In," *South China Morning Post*, December 26, 2014, www.scmp.com/news/hong-kong/article/1668692/burglars-steal -rare-turtles-worth-more-hk14-million-village-house.

8. "Malaysia Seizes 330 Smuggled Exotic Tortoises Worth $300,000," *Hindustan Times*, May 15, 2017, www.hindustantimes.com /world-news/malaysia-seizes-330-smuggled-exotic-tortoises-worth -300-000/story-yjBikgJzBtbn8QGkwBCMfI.html.

9. Tonya Strickland, "Mid-State Fair Game Operator Faces Fine over Baby Turtle Prizes," *Tribune*, July 16, 2013, www.sanluisobispo .com/entertainment/mid-state-fair/article39450924.html.

10. "TWRA Finds Turtles Given as Fair Prize, Returns Them to River," *WSMV.com*, August 25, 2011, www.wsmv.com/story/15333901 /turtle-with-possible-salmonella-sought-by-twra.

11. Kylen Mills, "People in Norman Upset after Baby Turtles and Rabbits Are Given as Prizes at the County Fair," *KOKH*, September 17, 2015, http://okcfox.com/news/local/people-in-norman-upset-after -baby-turtles-and-rabbits-are-given-as-prizes-at-the-county-fair.

12. "Salmonella and Turtle Safety," U.S. Food and Drug Administration, www.fda.gov/animalveterinary/guidancecomplianceenforce ment/complianceenforcement/ucm090573.htm.

13. "Take Care with Pet Reptiles and Amphibians," Centers for Disease Control and Prevention, April 3, 2017, www.cdc.gov/features /salmonellafrogturtle/index.html.

14. "The FDA Ban on Turtles: Fact and Fiction," Concordia Turtle Farm LLC., www.turtlefarms.com/FDA-BAN.htm.

15. Ibid.

16. Concordia partners with Petco for its Turtle Relinquish Program. Petco offers sanctuary for undersized or unwanted turtles at their stores, no questions asked. The company ships the abandoned turtles to Concordia for rehabilitation. Once there, Davey Evans and his crew "tend to 'em" and either sell them back to Petco once they grow to a legal size or use them for breeder stock in the pond. Animal rights advocates, such as People for the Ethical Treatment of Animals (PETA), criticize the recovery operation, claiming Petco should not be selling turtles and that Concordia is a "massive turtle meat factory."

17. Mark A. Rees, *Archaeology of Lousiana: Dispatches from the Modern South* (Baton Rouge: Louisiana State University Press, 2010).

18. As told by Ben Hirsch's daughter, Sondra Healy, and recounted by Gilbert White Prince of the Zeno Group public relations agency in a November 23, 2015, email to the author's research assistant.

19. "Our Story," Turtle Wax, www.turtlewax.com/about/history -heritage.

20. Prange Way, "Valley River Center, Eugene, Oregon." *Labelscar: The Retail History Blog*, October 27, 2010, www.labelscar.com/oregon /valley-river-center.

21. "City Facilities: Delta Ponds," Eugene, OR, website, www.eugene -or.gov/Facilities/Facility/Details/Delta-Ponds-133.

22. "Western Pond Turtle (Native)," Native Turtles of Oregon, www .oregonturtles.com/native_turtle.html.

23. Gregory McNamee and Luis Alberto Urrea, *A World of Turtles: A Literary Celebration* (Boulder, CO: Johnson Books, 1997).

24. Ani Vrabel, "On Overcoming a Debilitating Fear . . . of Turtles," *Huffington Post*, June 26, 2013, www.huffingtonpost.com/ani-vrabel /on-overcoming-a-debilitating-fear_b_3499809.html.

Chapter 6: The Illicit Hunts

1. U.S. Department of Justice, Office of Public Affairs, "Four Men Charged with the Illegal Trafficking of Threatened Alligator Snapping Turtles," *Justice News,* April 27, 2017, www.justice.gov/opa/pr /four-men-charged-illegal-trafficking-threatened-alligator-snapping -turtles.

2. U.S. Department of Justice, "Sentencings," *Environmental Crimes Section Monthly Bulletin*, January 2018, 16–17, www.justice.gov/enrd /page/file/1028911/download#January.

Chapter 7: The Wily Smugglers

1. Robert Snell, "Man Tried to Smuggle 51 Turtles in Pants Across Border," *Detroit News*, September 25, 2014, www.detroitnews.com/story /news/local/wayne-county/2014/09/25/man-tried-smuggle-reptiles -pants-across-detroit-windsor-border/16203929.

2. Robert Snell, "Feds Lift Veil on International Turtle Smuggling Ring," *Detroit News*, September 26, 2014. www.detroitnews.com/story /news/local/metro-detroit/2014/09/26/turtle-detroit-windsor-border -smuggling-china/16282617/.

3. Jeremy Blum, "Man Tries to Smuggle Turtle onto Plane by Hiding It in a Hamburger," *South China Morning Post*, July 31, 2013, www .scmp.com/news/china-insider/article/1293310/man-tries-smuggle -turtle-plane-hiding-it-hamburger.

4. U.S. Attorney's Office, Western District of Washington. "Snohomish County Man Who Smuggled Protected Reptiles Sentenced to Prison," United States Department of Justice, January 17, 2014, www .justice.gov/usao-wdwa/pr/snohomish-county-man-who-smuggled -protected-reptiles-sentenced-prison.

5. "Man Called Innocent Caught Smuggling Rare Turtles, Facing 10 Years in Jail," *Daily Star*, June 13, 2014, www.dailystar.co.uk/news /latest-news/383764/Man-called-Innocent-caught-smuggling-rare -turtles-facing-10-years-in-jail.

6. U.S. District Court, Western District of Washington, at Seattle indictment no. CR13 138MJP, May 2, 2013.

7. "Chapter 75. Endangered Species," Pennsylvania Code, www .pacode.com/seure/data/058/chapter75toc.html.

8. Mica Rosenberg, "Exotic Animals Trapped in Net of Drug Trade," *Reuters*, February 6, 2009, www.reuters.com/article/us-drugs-animals -odd-idUSTRE5154PM20090206.

9. U.S. Attorney's Office, Eastern District of Louisiana, "Illinois Man Sentenced for Violating the Lacey Act." U.S. Department of Justice, August 5, 2015, www.justice.gov/usao-edla/pr/illinois-man -sentenced-violating-lacey-act.

Chapter 8: The Frustrated Cops

1. A tip of Barry Baker's hat to Archie Carr, longtime University of Florida zoology professor and author of the 1956 classic natural history of sea turtles, *So Excellent a Fishe*.

2. D. H. Lawrence, *Tortoises*, (New York: Thomas Seltzer, 1921).

3. www.wwf-congobasin.org/where_we_work/gabon/gamba_com plex_programme.

4. "Shell Divests Gabon Offshore Interests," *Shell Global*, March 24, 2017, www.shell.com/media/news-and-media-releases/2017/shell-di vests-gabon-onshore-interests.html.

Chapter 9: The Pitiful Casualties

1. www.washingtonpost.com/news/the-fix/wp/2014/04/15/every thing-you-need-to-know-about-the-long-fight-between-cliven-bundy -and-the-federal-government/?utm_term=.57ba4aa90491.

2. www.nytimes.com/2014/04/24/us/politics/rancher-proudly-bre aks-the-law-becoming-a-hero-in-the-west.html.

3. Peter Pearsall, "Thunder the Rescued Olive Ridley Sea Turtle Has Died," joint news release, SeaWorld, U.S. Fish & Wildlife Service, and Oregon Coast Aquarium, April 13, 2016.

4. "Animals Head for Freedom As Argentina Closes Zoo," *Daily Mail Online*, July 2, 2016, www.dailymail.co.uk/wires/ap/article-3670995 /PICTURED-Animals-head-freedom-Argentina-closes-zoo.html.

5. Horacio Quiroga, *South American Jungle Tales*, trans. Arthur Livingston (New York: Dodd, Mead, 1941).

6. U.S. Department of Agriculture, Foreign Agriculture Service, "Dryness Continues in Argentina: Soybean Plantings May Increase,

with Corn Lower," *Commodity Intelligence Report,* January 14, 2014, https://ipad.fas.usda.gov/highlights/2014/01/Argentina/.

7. Ministerio de Agricultura, Ganadería y Pesca, *Argentina Lider Agroalimentario,* September 16, 2011.

8. International Union for Conservation of Nature, "*Chelonoidsis Chilensis* (Argentine Tortoise, Chaco Tortoise, Southern Wood Tortoise)," IUCN Red List of Endangered Species, www.iucnredlist.org/details/9007/0.

Chapter 10: The Conflicted Public and the Dedicated Conservationists

1. "A Streetcar Named Tortoise," *Time,* April 5, 1954, 45.

2. Donald Trump and Meredith McIver, *Never Give Up: How I Turned My Biggest Challenges into Success* (Hoboken, NJ: Wiley, 2008).

3. Deirdra Funcheon, "What Happened to Donald Trump's Threatened Tortoises?" *Fusion,* July 27, 2016, http://fusion.net/story/329068/what-happened-to-trump-tortoises/.

Chapter 11: The Imminent Future

1. Cahal Milmo, "Look Out! Abandoned Terrapins About," *Independent,* January 11, 2010, www.independent.co.uk/environment/nature/look-out-abandoned-terrapin-about-1863903.html.

2. "Illegal Turtles Seized at Airport," *BBC News,* November 14, 2008, http://news.bbc.co.uk/2/hi/uk_news/england/manchester/7728483.stm.

3. Harriet Hadfield, "Illegal Animal Item Seizures Rise by Millions," *Sky News,* November 14, 2013, https://news.sky.com/story/illegal-animal-item-seizures-rise-by-millions-10427889.

4. Anna Pukas, "Just What the Queen Always Wanted?" *Daily Express,* January 18, 2013, www.express.co.uk/news/royal/371516/Just-what-the-Queen-always-wanted.

5. Gilbert White, L. C. Miall, and W. Warde Fowler, *The Natural History and Antiquities of Selborne* (New York: Putnam, 1901).

SELECT BIBLIOGRAPHY

Benjamin, Walter. *Charles Baudelaire: A Lyric Poet in the Era of High Capitalism*. London: New Left Books, 1973.

Berolzheimer, Ruth. *The United States Regional Cook Book*. Chicago: Culinary Arts Institute, 1947.

Calloway, Stephen, and Katherine Sorrell. *Obsessions: Collectors and Their Passions*. London: Mitchell Beazley, 2004.

Filippini, Alessandro. *The Delmonico Cook Book: How to Buy Food, How to Cook It, and How to Serve It*. Bedford, MA: Applewood Books, 2008.

Geertz, Clifford. *The Interpretation of Cultures: Selected Essays*. New York: Basic Books, 1973.

Lawrence, D. H. *Tortoises*. New York: T. Seltzer, 1921.

McNamee, Gregory, and Luis Alberto Urrea. *A World of Turtles: A Literary Celebration*. Boulder, CO: Johnson Books, 1997.

Muensterberger, Werner. *Collecting: An Unruly Passion: Psychological Perspectives*. Princeton, NJ: Princeton University Press, 1994.

Quiroga, Horacio. *South American Jungle Tales*. Translated by Arthur Livingston. New York: Dodd, Mead, 1941.

Rees, Mark A. *Archaeology of Louisiana: Dispatches from the Modern South*. Baton Rouge: Louisiana State University Press, 2010.

Stanford, Craig B. *The Last Tortoise: A Tale of Extinction in Our Lifetime*. Cambridge, MA: Belknap Press of Harvard University Press, 2010.

Trump, Donald, and Meredith McIver. *Never Give Up: How I Turned My Biggest Challenges into Success*. Hoboken, NJ: Wiley, 2008.

White, Gilbert, L. C. Miall, and W. Warde Fowler. *The Natural History and Antiquities of Selborne*. New York: Putnam, 1901.

INDEX